Table of Contents

iv

List of Figures

AN ANALYSIS OF EYE MOVEMENTS WITH HELMET MOUNTED DISPLAYS

I. Introduction

1.1 General Issue

Advancements in technology have increased the quality and flexibility of Helmet Mounted Displays (HMDs), which has increased their application in operational environments. Many air and ground platforms use HMDs to maintain information in the user's field of view regardless of head position. The availability of information is appealing when quick decisions must be made, which encourages widespread HMD use. For example, the F-35 has replaced the Head-Up Display (HUD) with an HMD to display all critical flight information (1). Special operations groups use HMDs while riding All Terrain Vehicles or other land-based vehicles. The constant exposure to and availability of usable data improves user performance on tasks such as missile targeting, as well as improving user situation awareness (30).

Although HMDs have demonstrated improved user situation awareness in many situations, under low frequency vibration, 10 Hz and below, HMDs suffer from a reduction in effectiveness. Experts attribute this loss of effectiveness to the pilot's inability to perceive the information displayed while in motion to the human Vestibulo-Ocular Reflex (VOR). The VOR attempts to correct the eye position when the otoliths in the inner ear detect head movement by moving the eye in an equal magnitude and opposite direction of the motion. The goal of the VOR is to stabilize the line-of-sight on objects viewed by the user. Although the VOR is essential to human performance under normal conditions, while wearing an HMD, the display moves with the head, and therefore the VOR causes the eye to

1

rotate away from the information the user is attempting to view. This eye movement results in the information displayed on the HMD to appear blurred, reducing the user's ability to obtain the information from the display. Movement of the helmet with respect to the head, referred to as helmet slippage, is also a recorded occurrence during vibration (24). With HMDs, movement of the helmet results in display and image movement. Another factor of interest is the pursuit reflex (PR), which isa neural reflex that moves the eye in response to movement of the object being focused on (12). At lower frequencies, the visual system may be able to track objects on the retina, permitting the PR to maintain fixation on a slowly moving target. However, at higher frequencies, when the PR is not able to respond quickly enough to maintain focus on the image, the information displayed on the HMD will become blurred. The inability of the user to utilize the HMD to gain information in low-frequency vibratory environments degrades the advantages a HMD provides and can degrade user performance especially if the HMD provides critical information.

Although most fixed-wing aircraft do not experience low frequency vibration under normal flying conditions, buffeting can cause vibration that induce the VOR for short periods. Additionally, rotary-wing aircraft commonly experience vibration that are typically associated with the VOR. Even if the vibration levels that induce the VOR are uncommon and occur for short periods, the situations in which these conditions occur are crucial moments, such as combat turns and stalls, that determine the success of a mission. The user's ability to obtain critical information at these times to make decisions and perform tasks is essential. Surface irregularities are the major source of vibration in ground vehicles, causing low frequency vibration that could excite the VOR as well.

1.2 Problem Statement

Although research has documented user performance degradation while wearing a HMD in a vibration environment (19), minimal experimentation characterizing the VOR or PR has occurred. To correct these issues, a better understanding of how the eye moves while using a HMD at different vibration frequencies is required. The overall goal of this research is to analyze human reaction to low frequency vibration, including eye movements that occur while completing tasks on a HMD, to determine how to stabilize the HMD image on the retina. A possible solution to this issue is the use of two accelerometers, one located on the helmet, the other located on the seat. A processor attached to the HMD would utilize inputs from the accelerometers to correct the image placement on the HMD so the user would be able to see the information at all times. The goal of this thesis is to investigate specific questions that must be addressed to assess the feasibility of this concept or to suggest more viable concepts.

1.3 Research Objectives/Hypothesis

This research project focuses on characterizing the user's eye movements while the user completes visual tasks on a HMD and experiences sinusoidal vibration of known frequency and magnitude along the z-axis. Over the course of this research, the experimentation must address several sub-objectives before being able to achieve the eventual goal of improving user performance on a HMD. These objectives include:

- Obtain a baseline analysis of eye movements while performing basic visual tasks on a HMD in a simple vibration environment.

- Analyze experimental data to associate the VOR and PR effect with reference to vibration frequencies, vibration magnitude, acceleration levels, head position, and head orientation.

- Develop an initial predictive algorithm for eye position based on the analysis of the experimental data that can be used to correct image placement in HMDs.

- Analyze the VOR effect in a multi-axis, real-world vibration environment and compare the effects to the simple vibration environment data to determine the effectiveness of characterizing the VOR in a complex environment.

- Implement the predictive algorithm in a multi-axis, real-world vibration environment to analyze the feasibility and usefulness of a compensatory algorithm.

Results from previous vibration and VOR research, in addition to real-world performance reports, allow for several hypotheses about the results of this research to be made.

- It is possible to track eye movements in order to analyze eye movements in a vibration environment using electro-oculography (EOG).

- The VOR will be able to compensate effectively for head movement, except for in the 4-8 Hz range, where the gain and phase of the VOR will be unable to synchronize. At higher frequencies, the body dampens the vibration, increasing the VOR effectiveness.

- Due to more complex eye movements, the VOR will be less effective during a tracking task than a fixation task.

- Compensatory algorithms and associated hardware can be developed for HMDs after successful tracking of eye movements and analysis of VOR effects.

1.4 Research Focus

Based on the complexity of the eventual research goal, the focus of the current research is to characterize eye movements with respect to vibration frequency, acceleration level, head motion, and head position. Previous research by Uribe began a baseline analysis of VOR effects that will provide the basis of this research (27). By simplifying real-world environments to a single axis, and conducting simple visual tasks, the general eye movements can be identified.

1.5 Investigative Questions

In accord with the focus of this research, the questions designed to guide this study are:

- To what extent can eye movements be accurately tracked and characterized in a vibration environment?

- To what extent are eye movements predictable while undergoing vibration?

1.6 Methodology Overview

The method will include designing and conducting a human-subjects experiment, then applying data to answer the research questions. The experimental data will be obtained from human subjects wearing a modified helmet, undergoing low frequency, sinusoidal vibration along the z-axis. Previous studies have documented a range between 2-10 Hz as the frequencies that have the greatest effect on human biodynamics and performance. A modified HMD will present simple visual tasks, such as fixation and tracking, to the subject as well as an infrared diode and digital video camera to illuminate the eye and record video of eye movements. In addition to the video, EOG electrodes will be placed on subject's forehead and around the eyes to monitor eye movement with respect to the head.

1.7 Assumptions/Limitations

Many simplifying assumptions are necessary to develop the baseline data, as little research has been previously conducted in this area. Utilizing a single-axis vibration table to conduct the experiment allows the independent variables to be isolated, but is not representative of the complex vibration that users experience in real-world environments, which consists of vibration in multiple axes at a variety of frequencies. Additionally, only one vibration frequency will be applied at a time, to analyze the causes of the eye movements. Simple visual tasks will be completed, one at a time, to characterize the eye movements in various conditions, unlike operational conditions which require multiple tasks to be completed simultaneously, or with very little transition time. These limitations affect the realism of the experiment. It may be difficult to extrapolate the simplified conditions to the more complicated real world. In addition to the simplified environment, vibration effects vary among people based on several factors such as body type and posture, making general application of the research results challenging.

1.8 Implications

Although past research and development projects for HMDs have attempted to correct for vibration using compensation algorithms, employing various filters that utilize inputs from the environment, such as platform vibration, without understanding eye movements, these efforts have not been successful. Specifically, these endeavors to correct vibration have suffered from latency issues and inaccurate compensation for eye movements. In the absence of competent algorithms larger text/graphic have been implemented to increase user performance under all conditions. This reduces the potential effectiveness of HMDs by reducing the amount of information that can be displayed and counteracting the advancements in technology that have been made in high resolution HMDs. The eventual goal of this research is to generate an accurate compensation algorithm for eye movements. This algorithm would be integrated onto future HMDs to correct the display while the user experiences vibration. The stable image would reduce the required size of text and graphics displayed on the HMD, allowing more information to be presented to the user. This correction would allow the user to be able to perceive the information displayed under all conditions, increasing their situation awareness and ability to effectively complete their mission.

1.9   Summary

The academic format for this thesis was chosen to present the research and findings. Chapter 2 contains a paper accepted into the 2014 Industrial and Systems Engineering Research Conference (ISERC), authored by Kalyn Tung, Dr. Michael Miller, Dr. John Colombi, and Dr. Suzanne Smith. The paper contains the analysis of the fixation task data. Chapter 3 is a article to be submitted to the Society of Information Display (SID). The article describes the findings for Uribe's work and compares his results with the results from the experiment completed for this thesis. The article is authored by Kalyn Tung, Dr. Michael Miller, Daniel Uribe, Dr. John Colombi, and Dr. Suzanne Smith. Chapter 4 contains the conclusions from the two papers and recommendations for future work.

II. Eye movement in a Vibrating HMD Environment

Helmet or Head-Mounted Displays (HMD) applications have expanded their range from advanced military cockpits to consumer glasses. However, users have documented loss of legibility while undergoing vibration. Recent research indicates that undesirable eye movement is related to the vibration frequency a user experiences. In vibrating environments, two competing eye reflexes likely contribute to eye movements. The Vestibulo-Ocular Reflex (VOR) responds to motion sensed in the otoliths while the pursuit reflex is driven by the visual system to maintain the desired image on the fovea. This study attempts to isolate undesirable eye motions that occur while using a HMD by participants completing simple visual tasks while experiencing vertical vibration at frequencies between 0 and 10 Hz. Data collected on participants' head and helmet movements are compared to eye movements to develop a method to understand the source of the unintended eye movements. These experiments are needed to expand our understanding of vibration-induced eye movements, with the objective to construct systems, which ameliorate the effects of low frequency vibration on HMD legibility.

Keywords

Human Factors, Displays, Vibration Effects

2.1 Introduction

The use of the modern Head or Helmet Mounted Display (HMD) began in the 1970s as a method for off-boresight targeting on fighter aircraft (7). Advancements in HMD capabilities have expanded their use into many application areas, such as first-person video (FPV) radio-controlled aircraft, and recently high profile consumer products, such as Google Glass.

In spite of the enormous benefits HMDs provide, several documented human factors issues have persisted. As advancements in HMDs occur, the support systems become more complex. Excessive amounts of information being processed results in perceptible latency, a difference between an input signal and the appropriate response from the system. This can result in inaccurate information being displayed to the user at a crucial time (7). As HMDs are capable of presenting a rich set of information displaying too much information, or improperly displaying information can distract the user, forcing them to concentrate on the display, not their surroundings, to obtain information (26). Another issue is the fatigue associated with HMDs. The complex systems associated with the HMD add weight to the helmet. Excessive weight and unusual center of mass positions can cause muscle fatigue, especially at higher g-levels (17). Users have also documented issues with perceiving information displayed at low frequency vibration levels because of blurring (18). The effect of low frequency vibration on eye movements while wearing a HMD is the focus of this research.

2.1.1 Vibration.

Vibration can be transferred from the environment to the user and the user's head in many environments, but is prevalent in vehicles. For example, aircraft experience vibration along all axes at frequencies from 0.05-100 Hz. Fixed wing jet aircraft can experience inconsistent vibration, known as buffeting, normally during crucial flight regimes just prior to aircraft stall. The F-15 experiences sharp vertical accelerations around 8.5 Hz during tactical maneuvers, such as high angle of attack, high-g turns (24). Rotary wing aircraft experience more consistent vibration levels, with the main rotor as the source for the most prevalent frequencies, and the tail rotor producing secondary vibration (19). Vibration also occurs in ground-based vehicles when traveling on rough roads or other terrain (33).

Vision decrements caused by vibration are well documented. These decrements occur for HMD users as vibration is transmitted through the body to the head and are especially troubling at frequencies near the whole body resonance frequency (~4-8 Hz) where the highest head motions occur. Although the magnitude of the effects vary significantly based on frequency and acceleration levels, the vibration effects at frequencies of 15 Hz and below are primarily caused by body transmission (22). The transmission of low frequency vibration through the body is dependent upon many factors, such as posture, weight, and orientation.

Smith and Smith conducted an experiment analyzing the effects of head orientation on a tracking task while wearing a helmet. The participants were vibrated on a six-degrees-of-freedom vibration table while the head, helmet, and slippage accelerations were measured in addition to tracking errors and percent time on target. Tracking errors were largest and percent time on target were lowest when the head was oriented to the side, indicating off-axis head movements degrade visual performance in vibration environments (25).

2.1.2 Human Eye Reflexes.

The eye has multiple methods of adjusting to the environment to improve perceptual ability. These adjustments occur constantly, and usually without conscious effort. The two reflexes that are expected to contribute significantly to the eye movements while wearing a HMD in a vibrating environment are the Vestibulo-Ocular Reflex (VOR) and the Pursuit Reflex (PR). The VOR shifts the eyes in response to perceived motion from the vestibular system to compensate for motion of the user's head. This reflex is most noticeable during activities such as running or jumping, where the user's head is moving with respect to objects which are focused on and would be imperceptible without constant adjustments of the eyes to compensate for the movement of the user's head and has been documented to occur during vibration (20). The VOR is primarily effective in compensating for rotational movements of the head, as translational movements cause changes in the distance from the object to the eye, which cannot be determined by the vestibular system (12). The frequencies at which the VOR is effective at are still being investigated, with some research indicating the VOR operates up to 10 Hz (31), while others indicate up to 20 Hz (32). The PR adjusts eye position as the object of interest moves with respect to the observer. The eye's ability to smoothly move in pursuit of a target varies based on the velocity of the target, as well as the direction of movement. Generally, the PR is effective in compensating for relatively slow object motion but becomes less effective with higher object speeds.

While wearing a HMD, objects displayed on the screen move with the user as the HMD moves with their head. However, the vestibular system perceives the motion, either from the user moving their head, or from the environment exerting forces upon the user, and the VOR attempts to compensate for this motion. When wearing a head-fixed HMD, the VOR causes the eye to move away from the objects the user is attempting to view on the display. As a result, the image of information presented on the HMD moves on the user's fovea, resulting in image blur, degrading visual performance.

2.2   Motivating Research

Although the effect of vibration on visual performance, such as reading, has been studied extensively in the past (22; 12), recent research in our laboratory by Uribe has focused on better understanding eye movements as a function of vibration. During this research, participants completed simple visual tasks on a constructed HMD while undergoing low frequency vertical vibration. Uribe recorded video of the participants' right eye during the experiment using a microcamera attached to the helmet, while simultaneously employing electro-oculography (EOG) to determine eye movements, and an accelerometer to determine vertical head motion. The EOG results have been reported elsewhere (29; 28). The original goal of this research was to analyze both the video and EOG data to further understand the eye movements.

The video analysis demonstrated that not only was the eye moving in response to vibration, but the HMD-fixed camera was moving with respect to the user's head during vibration.

To understand the helmet motion, a program was created in MATLAB which correlated an image of the participant's eye while viewing a fixation target without vibration to images of the same participant's eye that was collected on the same target while undergoing vibration. Vertical offset corresponding to the peak correlation was recorded.

The root-mean-square (RMS) values from the vertical offset data were calculated and compared to the EOG RMS values Uribe developed from the same participants (27). Figure 2.1 presents a comparison of the calculated values.

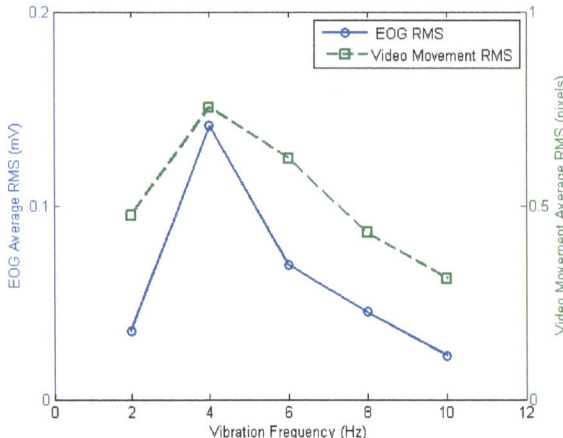

Figure 2.1. Comparison between EOG average RMS values and Video average RMS values

As shown, the shape of the curve created by mapping helmet movement as a function of frequency matches the EOG RMS values as a function of frequency. This analysis raises questions as to whether the eye movements as recorded using the EOG are created by the VOR in response to head movement, or in response to PR tracking movement of the helmet. Further it is unclear whether the VOR response and helmet motion are compensating for one another or combining to further degrade visual performance.

Unfortunately, the previous research did not record helmet rotational motion or slippage of the helmet on the user's head. Therefore, it was necessary to conduct a follow-on study, with the purpose of understanding head, helmet, and slippage motions, and how they relate to eye movement while wearing an HMD.

2.3   Methodology

The method involved a human-subjects experiment in which participants were exposed to low frequency vibration between 2 and 10 Hz at two different acceleration levels, including 0.1 g and 0.2 g. Each participant wore a helmet-mounted display which was equipped with a miniature video camera, an IR illumination source, and a set of accelerometers. Additionally, the participants were fit with a bite bar containing a different set of accelerometers and EOG leads to permit recording of eye movement data.

2.3.1   Participants.

Eleven participants between 22 and 29 years old volunteered to take part in the experiment, including 10 male participants. The participants were screened to not have any inner ear or eye injuries which could affect the eye movements of interest. The participants were not allowed to wear glasses during the experiment, however, soft contact lenses and corrective surgeries were allowed. Female participants were not allowed to be pregnant or have breast implants.

2.3.2 Vibration Table.

During the experiment, participants were seated upright on a human-rated vertical-axis servohydraulic vibration table. Participants were secured into a rigid seat with seat pan and seat back cushions, that was attached to the vibration table, similar to seats found in cockpits of fixed-wing aircraft, as shown in Figure 2.2. The facility technician ran the vibration table during the experiment, inputting the proper sinusoidal frequencies and displacement values necessary to attain the specified acceleration levels.

Figure 2.2. Subject on vibration table

2.3.3 Helmet.

A GENTEX HGU/53 Aircrew Helmet was selected, as it is a commonly used helmet onboard rotary and fixed-wing aircraft. The helmet consists of a hard, protective outer layer, a padded interior layer for comfort, and a chin strap to secure the helmet. It also has mounts for a visor on either side of the opening for the user's face.

A custom visor was created to fit into the mounts on the helmet, shown in Figures 2.3 and 2.4. The visor's main components were two adjustable side brackets and a plate across the front. The plate attached to a binocular LCD display system that was acquired from a pair of Vuzix Wrap 920 augmented reality glasses and provided 640 by 480 pixels over about a 31 degree diagonal field of view. The display system had been used in previous research to display simple visual tasks to users, simulating the types of task accomplished on a HMD. The plate also had a secondary purpose of blocking a majority of the ambient light, allowing the user to more easily view the display system. The adjustable sides allowed the user to adjust the distance to the display to improve the visibility of the display. The visor also included a bracket that held an IR LED and a SuperCircuits PC206XP micro camera, which were directed at the user's right eye. The power of the LED was adjustable up to a maximum level which was approximately one fifth the eye safe level for the illumination distance applied during the experiment.

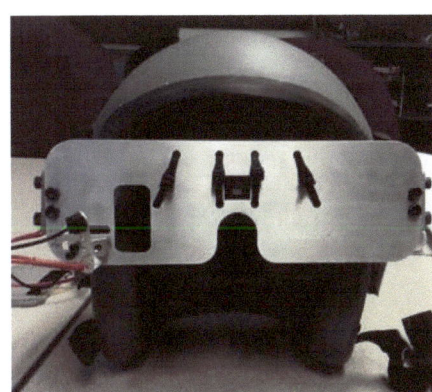 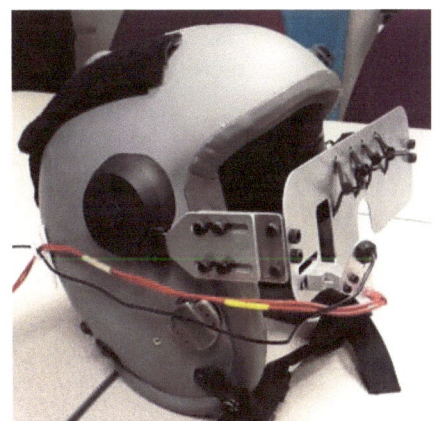

Figure 2.3. Modified helmet from the front.   Figure 2.4. Modified helmet from the side.

2.3.4 Electrooculography.

A BIOPAC MP150 data acquisition system was used to collect the EOG signal. EOG was applied to measure the potential difference, in mV, between electrodes on opposing sides of the eye. The Acknowledge software from BIOPAC was used to provide a vertical movement and a horizontal movement signal at a sampling rate of 1000 Hz.

2.3.5 Accelerometers.

A triaxial accelerometer pack was used to measure the input acceleration at the base of the seat in the fore-and-aft (X), lateral (Y), and vertical (Z) directions. The pack consisted of three-orthogonally-arranged miniature accelerometers (Entran EGAX-25) embedded into a two-piece plastic disk. Two sets of six miniature accelerometers (Entran EGA 125-10D) were orthogonally-arranged and strategically glued inside of a two-piece plastic mount, and attached to a bite bar and to the top of the helmet. This accelerometer arrangement permitted the measurement of the translational and rotational movements of the user's head and helmet. The bite bar was custom made for each participant's teeth, obtained using dental impression compound and a mouthpiece for a secure fit. This custom fit bite bar permitted accurate accelerometer readings of the participants head motions.

All acceleration data were collected onto a 48 channel data acquisition system (EME, Corp), filtered at 250 Hz, and digitized at 1024 samples per second. The EOG and accelerometer data were triggered simultaneously for each 15 seconds of task completion at each frequency.

2.3.6 Experimental Procedure.

Upon arriving, participants completed an informed consent and underwent appropriate screening procedures. The participants were then seated on the acceleration table and instrumented as necessary. The participants then completed the first of two test sessions.

During each test session, the participants first completed an eye movement calibration. The calibration required the participants to fixate on a target at a sequence of five locations, indicating the center and the extremes of target location during the experiment. The measurements from the calibration period were used to understand the magnitude of the EOG signals for each participant. The participants then completed three visual tasks at each frequency and at a fixed acceleration. The first task was a fixation task, during which the participant focused on a dot which remained stationary in the middle of the screen. The target used in the task was a black dot with a strong contrast to the background, sized so it would be easily located, but not so large that the participants had a large area to fixate. The participants also completed both a horizontal tracking task and a target search task. However, only results from the fixation task are reported in this paper. Upon completing the tasks at one acceleration level, the instrumentation was removed from the participant; they were debriefed and permitted to leave. Participants then completed the second test session on a different day, following the same procedure, only experiencing a different acceleration level.

2.3.7 Experimental Design.

The participants were exposed to six different sinusoidal vibration frequencies, including 0, 2, 4, 6, 8, and 10 Hz. These frequencies were experienced at 0.1 g ($\sim$0.69 m/s^2 rms) peak acceleration level for one test session and 0.2 g ($\sim$1.37 m/s^2 rms) peak acceleration level for the other test session. The order the acceleration levels were experienced by the participants was counterbalanced. These vibration levels and their durations were compared to the current human exposure guidelines (ISO 2631-1: 1997), with overall levels indicating there was low risk of health effects. The procedure was reviewed and approved through the Air Force IRB process.

As the task was a simple visual staring task, no learning effects were expected. Previous research indicates that the largest eye movements due to the VOR were expected at frequencies in the 4-6 Hz range, and as each frequency included a 15 second acclimation period, the effect of the order of the frequencies was expected to be minimal.

2.3.8   Data Analysis.

The EOG data for the fixation task was analyzed by first decomposing the signal into trial portions and selecting the longest time period in each without a blink. Blinks were clearly indicated by a very large amplitude spike in the vertical EOG signal. The EOG signal was then normalized by applying a normalization constant for each participant. The normalization constant was determined by applying a 0.5 s moving average to the portion of the calibration signal corresponding to the movement of the eye from the top to the bottom of the display. The 90th percentile value from the segment when the participant was looking toward the top of the screen, and the 10th percentile value from when the participant was looking towards the bottom of the screen were obtained. The difference between the 90th and 10th percentile values was estimated to represent the magnitude of the EOG signal which corresponded to the movement of the eye from the top to the bottom of the display.

A 0.5 second moving average was calculated and subtracted from the normalized signal to act as a high pass filter, removing low frequency drift. This centered the EOG signal on 0, and the remaining signal was assumed to represent eye movement in response to the vibration.

Power Spectral Densities (PSDs) for each of the signals were calculated. While most signals demonstrated a peak in the PSD at the applied frequency, in some cases, other frequencies had a higher peak, and in a few cases, no obvious peak at the applied frequency was shown. As a result, the RMS value of the EOG signal was calculated to capture the magnitude of the eye movements at all frequencies, rather than only at the applied frequency. The RMS values were used as an estimate of the eye movement at the respective frequency and acceleration level. The EOG signal RMS values for one subject were almost 3 standard deviations from the mean in 3 of the 5 frequencies during one test session; the large influence of these values resulted in that subject's data being excluded from further analysis.

Helmet slippage was calculated as the difference between the time histories of the bite bar and the helmet in each of the corresponding directions, X, Y, and Z. The rotational accelerations were calculated based on the subtraction of the time histories between the associated accelerometers, and then dividing by the moment arm. Rotational displacements were estimated using the FFT of the acceleration signal, dividing by the square of the angular frequency, and multiplying by -1. The inverse FFT was then applied to yield the estimated displacement time history. The acceleration and displacement RMS spectra were estimated from the respective PSDs in 0.5 Hz increments. Since the vibration was limited to the vertical axis, only the vertical translation accelerations and pitch rotational accelerations and displacements were used in the analysis. The RMS values associated with the input frequency were extracted from the spectra.

A repeated measures analysis of variance (ANOVA) was conducted on the RMS values from both the EOG and the accelerometers to understand the effects of the different frequencies and acceleration levels. As the data consistently failed Mauchley's test of sphericity, the Greenhouse-Geisser correction was consistently applied across each reported measure. Schefe's Least Significant Difference (sLSD) post-hoc analyses were applied to determine differences between frequency levels. A significance level of 0.05 was used during the statistical analysis.

19

2.4 Results

Figure 2.5 shows the mean RMS values plus and minus one standard deviation of the EOG signal at the applied frequencies and acceleration levels. As can be seen the EOG RMS values appear higher at 4 and 6 Hz than at the other frequencies as expected. The ANOVA indicated that frequency had a significant effect on the EOG RMS values ($F_{(1.85, 16.62)}=5.95$; $p=0.013$)). The effect of the acceleration level approached significance ($F_{(1,9)}=3.67$; $p=0.088$). The interaction of frequency with acceleration was not significant. The pair-wise comparisons indicated that the RMS values at 4 and 6 Hz were significantly higher as compared to the 2, 8 and 10 Hz conditions at both acceleration levels.

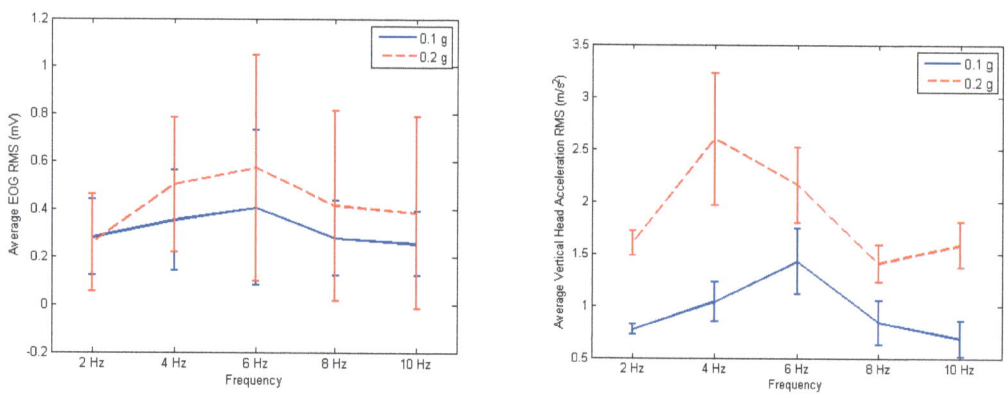

Figure 2.5. Mean and standard deviations of the Normalized EOG RMS values at different frequencies

Figure 2.6. Mean and standard deviations of vertical head acceleration RMS values at different frequencies

Figure 2.6 presents the mean RMS values plus and minus one standard deviation for the vertical head acceleration at both acceleration levels. The ANOVA indicated that frequency had a significant effect on the vertical head acceleration ($F_{(2.11,19.02)}=25.01$; $p\leq0.000$). The effect of acceleration level was also significant ($F_{(1,9)}=568.359$; $p\leq0.000$), as was the interaction between frequency and acceleration level (F(1.43,12.91)=9.32; p=0.005). To analyze the interaction the ANOVA was applied for each acceleration level. At 0.1g, the RMS values for 6 Hz were significantly larger than all other frequencies. At 0.2 g, the RMS values for both 4 and 6 Hz were significantly larger than the RMS values at 2, 8, and 10 Hz.

Figure 2.7 shows the mean RMS values plus and minus one standard deviation for the head pitch acceleration as a function of frequency at both acceleration levels. For the head pitch acceleration, as with the vertical head acceleration, the effects of frequency (F(2.23,20.10)=26.39; p≤0.000), acceleration (F(1,9)=21.636; p=0.001) and the interaction (F(2.10,18.91)=3.713; p=0.042) were significant. At 0.1 g, the RMS value at 4 Hz was significantly larger than for 2 and 10. The RMS value at 6 Hz is significantly larger than the RMS values for all other frequencies. At 0.2 g, the RMS value at 6 Hz was significantly larger than the RMS value at 2, 8 and 10 Hz. The RMS value for 8 Hz was significantly different from the RMS values for 2, 4, and 6 Hz.

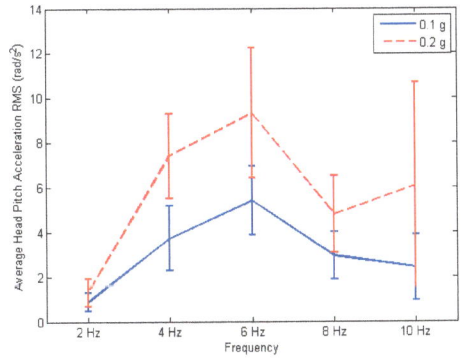 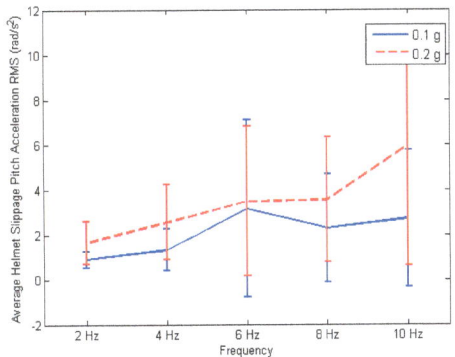

Figure 2.7. Mean and standard deviations of head pitch acceleration RMS values at different frequencies

Figure 2.8. Mean and standard deviations of helmet slippage pitch acceleration RMS values at different frequencies

Figure 2.8 shows the mean values plus and minus one standard deviation for the helmet slippage pitch acceleration. The effect of frequency ($F(1.82,16,41)=5.52$; $=0.017$) was significant. Acceleration ($F(1,9)=3.09$; $p=0.113$) and the interaction term ($F(2.09, 18.85)=1.53$; $p=0.241$) were not significant. The average helmet slippage pitch acceleration RMS values generally increased as frequency increased. Only 2 Hz was significantly different from all other frequencies.

Figure 2.9 shows the mean RMS values plus and minus one standard deviation for the head pitch angle. Frequency ($F(1.88,16.89)=45.86$; $p \leq 0.000$), acceleration ($F(1,9)=36.70$; $p \leq 0.000$) and the interaction ($F(2.39,21.55)=7.92$; $p=0.002$) were significant. The ANOVA applied for each acceleration level indicated that the RMS values at 2 and 4 Hz were significantly larger than the RMS values at 6, 8 and 10 Hz. A similar trend existed at 0.2 g.

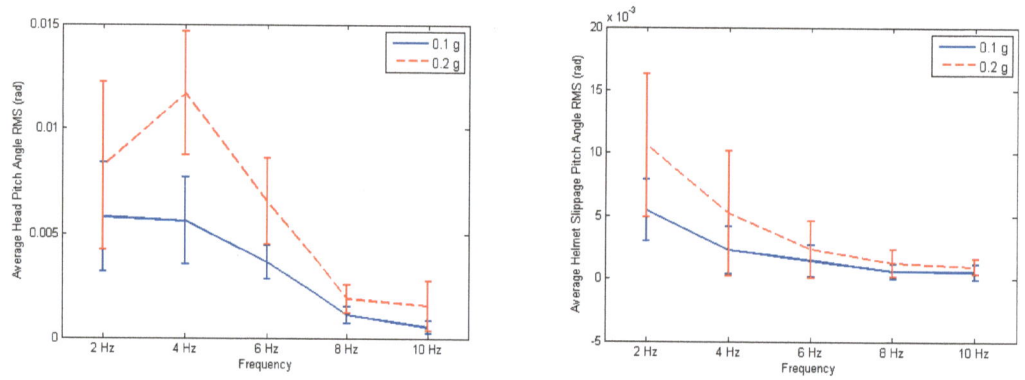

Figure 2.9. Mean and standard deviations of head pitch angle RMS values at different frequencies

Figure 2.10. Mean and standard deviations of helmet slippage pitch angle RMS value- sat different frequencies

Figure 2.10 shows the mean RMS values plus and minus one standard deviation for the helmet slippage pitch angle at both acceleration levels. The frequency ($F(2.12$,

19.12)=22.10; p≤0.000), the acceleration (F(1, 9)=21.40; p=0.001), and the interaction between frequency and acceleration (F(1.78, 16.04)=3.88; p=0.046) were significant. The ANOVA applied to each acceleration level indicated that at 0.1 g, 2 Hz was significantly larger than all other frequencies. Both 4 and 6 Hz were significantly different from 2, 8, and 10 Hz, and 8 and 10 Hz were significantly less than 2, 4, and 6 Hz. At 0.2 g, only 2 Hz was significantly larger than all other frequencies.

2.4.1 Discussion.

The primary goal of this research was to understand the relationships between head and helmet motions, including helmet slippage, and eye movements during low frequency sinusoidal vibration.

The results of this study showed that eye movements, characterized through the use of EOG signals showed the highest peaks at 4 and 6 Hz, and the magnitude of the responses were not statistically significant for the acceleration level. Comparing these results to the head and helmet motions, both the vertical head acceleration RMS and head pitch acceleration RMS showed curves that were similar in shape to the EOG curves. However, the highest peak occurred at 6 Hz at the lower acceleration level, and at both 4 and 6 Hz at the higher acceleration, indicating a dependence on the acceleration level. The significant effect of acceleration on the accelerometer values, but the lack of significant effect from acceleration level on the EOG values possibly indicates a higher noise level in the EOG signal, or the VOR response is not sensitive enough to perceive the changes in acceleration level.

While helmet slippage acceleration might have an effect on eye movements as the frequency and acceleration level change, the relationship is less obvious. For 2, 4, and 6 Hz the helmet slippage pitch accelerations increase similarly to the EOG values, however, at 8 and 10 Hz, the RMS curves appear to differ significantly. This might be caused by the eye's

23

reaction time. At higher frequencies the eye might not be able to register the movements and react appropriately. The similar values between the head pitch acceleration and the helmet slippage pitch acceleration values indicate that helmet slippage could have an affect on eye movements.

The helmet slippage pitch angle RMS indicates the amount the helmet, and therefore the display, moved with respect to the head. Although the motivating research indicated there might be some relationship between the EOG signal and the helmet slippage, the relationship does not appear to be as significant as hypothesized once more detailed values were produced.

To further explore the interrelationship of the variables, a stepwise regression to predict the EOG RMS values for individual participants and experimental conditions was conducted using frequency, acceleration level, and the accelerometer values as inputs. The resulting function included the head pitch acceleration and helmet slippage pitch acceleration value as significant factors and accounted for 28.3% of the variability in the data. The estimated regression function is shown in Equation 2.1. The estimated regression function

$$EOG = 0.3399 + 0.0363 \cdot HeadPitchAcceleration - 0.0473 \cdot HelmetSlippagePitchAcceleration$$

$$(2.1)$$

The regression function indicates that the head pitch acceleration and helmet slippage pitch acceleration have relationships with the EOG signal. As the change in frequency had a significant effect on the EOG RMS values, this also indicates that any frequency effects are contained in the head pitch acceleration or helmet slippage pitch acceleration variables. Although the adjusted R^2 is a relatively low value, the large variance in the EOG RMS values between participants would be difficult to account for with this method.

2.5   Conclusion

Eye movements recorded through EOG indicate that eye movement magnitude is not linearly related to the vibration input to a seat pan. This finding provides one potential reason why previous algorithms to compensate for vibration in HMDs have failed as these methods have attempted to compensate for the VOR as a function of vibration of the vehicle.

Results demonstrate that both head pitch angle and vertical acceleration are also not linearly related to seat pan vibration and appear to have a shape similar to the shape of the eye movement magnitudes as a function of frequency. Therefore, it may be possible to improve the accuracy of compensation algorithms, if these algorithms were designed to respond directly to head acceleration, without being concerned with the vibration of the aircraft.

Although it is possible that both the VOR and eye movements are also affected by helmet slippage with respect to the user's head as the user's visual system attempts eye movements to compensate for the relative motion of the HMD, the results of this study are not yet conclusive. Future research should focus on minimizing the helmet slippage, possibly through the use of a mask or padding, to isolate the effect of vibration on the eye movements, or using other methods to isolate the effects of the VOR from the PR.

In this study we captured EOG as a source for tracking eye movements. This method does not provide ideal tracking of the human eye, with the resulting signal containing significant noise and apparent drift, which precludes the determination of absolute eye position. Unfortunately, optical tracking of the eye with a helmet-mounted camera and illumination source is also complicated by the fact that the resulting video contains not only changes due to eye movements, but changes in eye location due to helmet slippage. Future research into the phenomena of interest in this study would benefit from the application of other eye tracking technologies which are more reliable in a vibrating environment.

Bibliography

[1] Bayer, M., C. Rash, and J. Brindle, "Introduction to Helmet-Mounted Displays", appears in *Helmet-Mounted Displays: Sensation, Perception and Cognition Issues*, C. Rash, M. Russo, T. Letowski, and E. Schmeisser, Eds., Fort Rucker, AL: U.S. Army Research Laboratory, 2009, 47-103.

[2] Stuart, G. W., McAnally, K. I., and Meehan, J. W., 2005, "2 Head-up displays and visual attention: integrating data and theory," Contemporary Issues In Human Factors And Aviation Safety, 25.

[3] Parr, J. C., Miller, M. E., Pellettiere, J. A., and Erich, R. A., 2013, "Neck Injury Criteria Formulation and Injury Risk Curves for the Ejection Environment: A Pilot Study," Aviation, Space, and Environmental Medicine, 84(12), 1240-1248.

[4] Rash, C., Hiatt, K., Wildsunas, R., Caldwell, J., Kalich, M., Lang, G., King, R., and Noback, R., 2009, "Perceptual and Cognitive Effects Due to Operational Factors", appears in *Helmet-Mounted Displays: Sensation, Perception and Cognition Issues*, Rash, C. Russo, M., Letowski, T. and Schmeisser, E. (Eds), U.S. Army Research Laboratory, Fort Rucker, AL, 675-795.

[5] Smith, S. D., 2002,"Collection and characterization of pilot and cockpit buffet vibration in the F-15 aircraft." SAFE Journal, 30(3), 208-218.

[6] Rash, C., McLean W., Mora J., and Ledford M., "Design Issues for Helmet Mounted Display Systems for Rotary-Wing Aviation," U.S. Army Aeromedical Research Laboratory, Fort Rucker, AL, 1998.

[7] Wong, J. Y., 2001,*Theory of ground vehicles*, Wiley.com.

[8] Shoenberger, R. W., 1972, "Human Response to Whole-Body Vibration." Perceptual and motor skills, 34(1), 127-160.

[9] Smith, S. D., and Smith, J. A., 2006, "Head and helmet biodynamics and tracking performance in vibration environments." Aviation, space, and environmental medicine, 77(4), 388-397.

[10] Roy, J. E., and Cullen, K. E., 1998. "A neural correlate for vestibulo-ocular reflex suppression during voluntary eyehead gaze shifts." Nature neuroscience, 1(5), 404-410.

[11] Griffin, M., 1994, *Handbook of Human Vibration*. 2nd Edition, Academic Press, San Diego.

[12] Vercher, J. L., Gauthier, G. M., Marchetti, E., Mandelbrojt, P., and Ebihara, Y., 1984, "Origin of eye movements induced by high frequency rotation of the head," Aviation, space, and environmental medicine, 55(11), 1046.

[13] Wells1983 Wells, M. J., and Griffin, M. J., 1983, "Vibration-induced eye motion," In Aerospace Medical Association Annual Scientific Meeting, Houston, Texas, pp. 23-26.

[14] Uribe, D. J., Miller, M. E., and Smith, S., 2013, "An analysis of the vestibule-ocular reflex during vibration," Proceedings of the 17th International Symposium on Aviation Psychology, 17.

[15] Uribe, D. J., and Miller, M. E., 2013, "Eye Movements When Viewing a HMD Under Vibration," Appears in Proceedings of the Human Factors and Ergonomics Society Annual Meeting, Vol. 57, No. 1, pp. 1139-1143, SAGE Publications.

[16] Uribe, D., 2013, "An investigation and analysis of the vestibulo-ocular reflex (VOR) in a vibration environment.," Masters thesis, Air Force Institute of Technology.

III. Comparing Eye Movements While Wearing a Helmet-Mounted Display

3.1   Introduction

Head or Helmet Mounted Displays (HMDs) were originally applied to aid fighter pilots aim at enemy aircraft by aiming weapons through mechanically following head position (7). Modern day HMDs have increased their capabilities to display critical information to the user, regardless of head position. In addition to aviation related applications, HMDs have spread into other environments, such as on-board land vehicles and commercial use (3; 2; 4). These displays permit information to be constantly available, permitting users to make decisions quicker based on improved situation awareness (10).

However several issues with HMDs have been documented. Attention capture occurs when the user focuses on the images displayed rather than their environment, thereby losing their awareness of their surroundings (26). Fatigue can result from increased head born mass (17). Degraded visual performance has also been reported for HMD used under vibration (18). This degradation of visual performance in the presence of vibration is of interest in the current research.

Two eye reflexes are expected to contribute to eye movements during whole-body vibration, the Vestibulo-Ocular Reflex (VOR) and the pursuit reflex (PR) (12). The VOR is triggered by the otoliths in the inner ear triggering eye movements to compensate for head motion. The presence of this reflex is most noticeable during high impact activities such as running or jumping as it permits the runner's eyes to remain focused on an object despite the vertical movement of the head during these activities. The PR is a neural reflex which guides the motion of the eye to follow the motion of an object being focused on when the object moves on the eye's fovea. Research indicates the pursuit reflex is able to suppress the VOR at low frequencies, such as 2 Hz and below (12), but is unclear on the relationship between these reflexes at higher frequencies.

Vibration can occur in many environments, produced by blade frequencies or buffeting in air vehicles, propulsion over rough surfaces in surface vehicles, or simple body motion during locomotion. However, the eye does not necessarily respond to the motion of the vehicle, but the motion of the observer's head. High head accelerations are most likely to occur near whole body resonance, which is accepted to be near 4 Hz for an individual seated upright (22). This finding is supported by Smith and Smith who investigated the effect of head orientation on head motions and tracking performance during vibration. This research demonstrated that the most extreme head and helmet rotations occurred in the pitch direction regardless of head orientation, with peak vibration levels between 2 and 6 Hz (25).

Research indicates the VOR is effective only at low frequencies, however specific ranges vary, some indicating up to 10 Hz, while others suggest the VOR is effective up to 20 Hz (12). The VOR is also only useful to counter rotational movements, as translational head movements result in changes in distance that the VOR cannot correct for (12).

Various methods have been attempted to counter the negative effects vibration has on HMD legibility. It has been demonstrated that by increasing the size, contrast, and spacing of the information displayed on the HMD visual performance can be enhanced (13; 11; 16). However, increasing the size of the text or images displayed to the user limits the amount of information that can be displayed. This compensation method also counteracts the advancements in screen resolution, which permit smaller objects to be displayed while retaining their clarity.

Another method to improve visual performance with a HMD while under vibration is to apply a filter to counter the vibration. Several compensation filters have been researched that adjust the location of the image based on the vibration input. However, they have not experienced consistent success at improving the user's visual performance, and have not been implemented onto any existing systems. Wells and Griffin attempted an image stabilization filter that countered all head motions using accelerometers attached to the helmet; however, the filter experienced issues during intentional head motions, causing movements in the displayed image when the movements were unnecessary (32). Daetz applied an adaptive filter to predict head vibration based on accelerometers attached to the aircraft, but these filters experience latency issues, given the complex transfer function necessary to quickly compute the expected output (8). Daetz also implemented a conventional notch filter to remove the noise due to vibration. While the performance on a tracking task improved, the filter did not compensate the image appropriately for other visual tasks (8). Another method to improve visual performance is the use of strobing the display. When vibrating at 0.7 g at 12 Hz, flashing a display in synchrony with the vibration reduced the reading error to a rate not significantly different from the non-vibrating error rate (5). The versatility of the HMD is difficult to account for, and an individual compensation filter has been unable to accurately adjust the display to allow for improved visual performance in all areas.

The original intent of this research project was to understand how the eyes move while using a HMD in a simple vibration environment. By understanding how the eye moves, it is hoped that it will be possible to design and build improved image compensation means for HMDs within a vibrating environment. A pair of experiments are reported to relate eye movements to the vibration, as well as the head and helmet movements experienced.

3.2   Experiment 1

The initial experiment was designed to understand the effect of various sinusoidal frequencies of vibration upon eye movements. As it was believed that eye movements might be influenced by the type of task the user performed, the experiment also required participants to perform two separate tasks, specifically fixation on a stationary target and tracking a horizontally moving target. Eye movements were recorded using multiple approaches as the advantages or disadvantages of these approaches were uncertain.

3.2.1 Experimental Methodology.

3.2.1.1 Participants.

Six volunteers (including 3 male) between the ages of 20 and 26 years with a mean of 23 years participated in the study. The participants had not experienced any vestibular anomalies, including inner ear infections, within the month preceding the investigation or reported any discomfort or pain symptoms associated with the musculoskeletal system. Individuals requiring glasses or hard contacts were precluded from participating in the experiment to simplify vision system-based eye tracking.

3.2.1.2 Apparatus.

During the experiment, participants were seated on a single-axis, vertical vibration table which imparted between 2 and 10 Hz vibration at approximately 0.1g to the seat on which they were seated. Participants were instrumented with an accelerometer mounted on an individually-molded mouth piece to measure head acceleration, electrodes to facilitate electro-oculographic recording of eye movements, and a helmet which provided both a display to provide a visual stimulus as well as a miniature video camera for eye recording and an IR illumination source to illuminate the eye, providing both a well-lit region around the eye and a specular highlight to potentially provide eye tracking information. This apparatus permitted a controlled sinusoidal vibration to be input to the participant and to record head accelerations and eye movements in response to the vibration. The low levels of vibration and brief exposures used in this study were associated with a low potential for health risk in accordance with the current human exposure guidelines (14).

31

The vibration table was located in a Single-Axis Servo-hydraulic Vibration Facility supported by the Air Force Research Laboratory's 711th Human Performance Wing. The table is shown in Figure 3.1. The table imparted vibration to the participants' chair. The human-rated single-axis vibration table is capable of recreating operational exposures in the vertical direction. A rigid seat with seat pan and seat back cushions were mounted on top of the Single-Axis platform. This seat was designed to transfer the vibration to the participant. Although the apparatus is capable of more complex vibration, only single frequency, sinusoidal signals were applied during the experiment. The sinusoidal frequency varied in 2 Hz increments between 0 and 10 Hz with estimated amplitude for a target acceleration level of 0.1 g. The table was not calibrated for each individual and therefore this amplitude likely varied somewhat as a function of the mass of the participant.

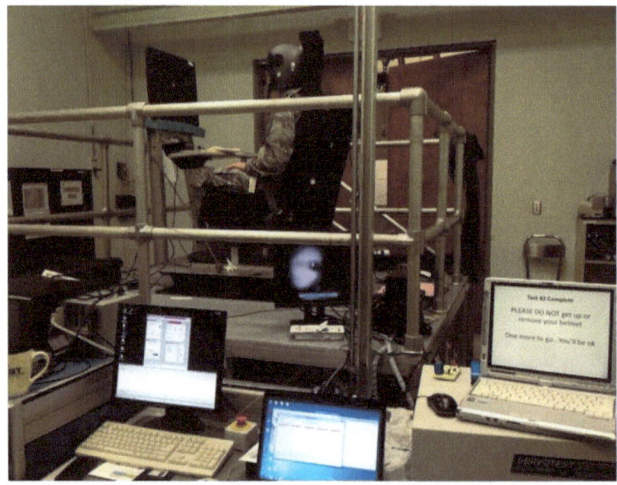

Figure 3.1. Single-Axis Servo-hydraulic vibration table.

A custom fit, gel mouth guard was molded for each individual participant prior to the experiment and a SS27 Tri-axial Accelerometer from Biopac was mounted on the mouth guard to measure the linear acceleration of the participants' head. Additionally, six electrodes were mounted on the participants' face to monitor the movement of the eyes with

32

respect to the head via Electro-Ogulography (EOG). A BioPac MP150 recorded both the linear acceleration and EOG signals with a sample rate of 1000 Hz. This EOG procedure recorded the potential difference between the electrodes as the eye moved from the center, neutral position towards the electrodes, which were positioned above and below the right eye and to either side of the eyes to obtain a vertical and a horizontal eye movement signal, respectively.

The participants were then fit with a HMD system, which supported visual tasks while blocking the participant's vision of the environment in their central visual field. The HMD included a typical Air Force Flight Helmet, equipped with a tinted visor. The helmet designed for this experiment is shown in Figures 3.2 and 3.3. A 640 by 480 pixel color, binocular LCD display was mounted on the outside of the visor. This display was taken from a pair of Vuzix Wrap 920 augmented reality glasses and was driven by a laptop computer to provide an image to each eye with a field of view of 30 degrees and a refresh rate of 60 Hz. The display included optics to provide an image at optical infinity. A microcamera (SuperCircuits micro-lens video camera) was attached to the helmet and focused on the eye. This camera is 0.375 in x 0.675 in in size and outputs sufficient image quality for analysis with 420 TV lines. This camera was connected to a digital video recorder to permit a video of the user's eye to be recorded throughout the experiment. A visibly opaque # 87 Kodak Wratten 2 Optical Filter was fit in the barrel of the camera to permit the camera to receive energy only in the IR portion of the spectrum. Additionally, an infrared LED was mounted about 2 inches from the camera and illuminated the participant's right eye. This LED both illuminated the eye and the region of the participants' face around the eye as well as provided a specular highlight from the back of the cornea, making it possible to use this feature for future analysis of the images to determine eye orientation. Custom electronics were used to drive the LED and provide adjustment. These electronics insured a safe level of exposure, permitting a maximum exposure of 1.94 mW/cm^2 at the eye, but was adjusted to provide only enough illumination to satisfactorily illuminate the eye prior to each condition. The maximum value is well below the maximum safe value of 10 mW/cm^2 for constant IR exposure in the 720-1400 nm range (15). The design of the miniature camera and illumination source was adapted from a design by (6).

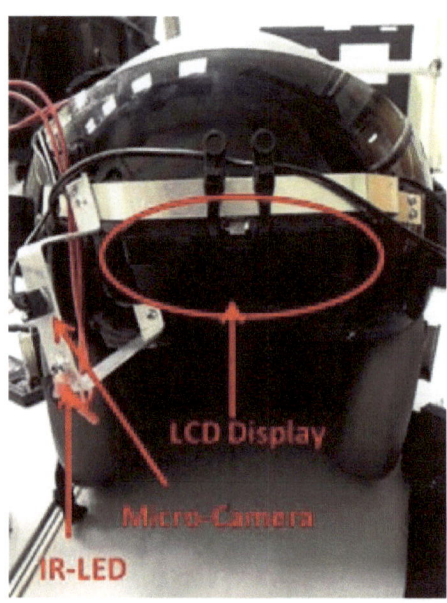

Figure 3.2. Helmet design from Experiment 1 - front

3.2.2   Experimental Procedure.

During the experiment, participants completed two tasks. Each task required the participant to fixate on a cross-hair target. During the first task (Task A), the target was stationary in the center of the field of view. During the second task (Task B), the same target moved around the display in a rectilinear pattern. During Task B, the target began at the center of the display, moved horizontally to the left edge of the display and followed the perimeter of the display. This target increased in velocity from each apex in the pattern, reaching a

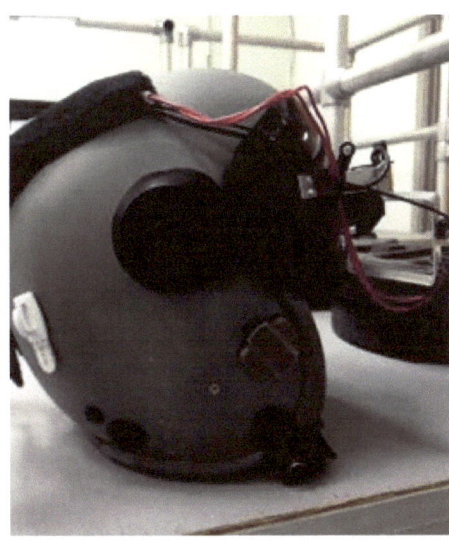

Figure 3.3. Helmet design from Experiment 1 - side

rate of 190 pixels per second and maintained this rate before decelerating to zero velocity as it approached the next apex. This velocity profile permitted the participants to track the target without overshooting as the target changed direction. Participants completed two trials, performing each task at each vibration frequency on two separate days. On the first trial, the vibration frequency increased with each sequential exposure while the frequency order was randomized for the second trial.

During each experimental trial, each participant first viewed a set of calibration targets on the display which were displayed at the center and the corners of the display. Each participant was then exposed to a vibration condition for 15 seconds for acclimation and then performed the task under the same vibration for 15 additional seconds. Once completed, the frequency of vibration was changed and the task repeated for the subsequent vibration condition. The order of frequency conditions differed between the first day's trial and the second. The participants completed both Task A and Task B on each of the two days, with half completing Task A before Task B. Given this procedure, the factors of Task

(A or B), Frequency (0, 2, 4, 6, 8, 10 Hz) and Trial (Day 1 or 2) were combined in a full factorial experimental design.

3.2.3   Data Analysis.

The primary dependent variable of interest was the magnitude of vertical eye movement, indicated by a change in the vertical EOG signal measured in mV. An additional analysis investigated the consistency of head acceleration.

The data analysis applied the following steps to the vertical component of the EOG data. First, the program segmented the signal collected during the fixation in Task A or horizontal target tracking Task B from the remainder of the data. The program applied a moving average filter to the segmented data to eliminate low-frequency drift in the EOG signal by applying a low-pass filter. The program then applied a second filter to determine high amplitude values caused by blinks or other erroneous eye movements, removing this data from the data stream. The program then determined the Root Mean Square (RMS) value in mV for each participant at each frequency condition and each task during each trial. The program then calculated the power spectrum using Welch's method (9), estimating the power spectral density (PSD) at different frequencies using an overlapping window principle and computing the discrete Fourier Transform. The resulting array was the RMS value as a function of frequency for each task. Finally, the program selected the peak frequency. This frequency consistently corresponded to the frequency of the input signal for all frequencies between 2 and 10 Hz. Unfortunately, one participant's data for one trial differed from all other data by more than an order of magnitude. Examination of video recordings of the participant's eye did not appear different from the video recording for any other trial. This discrepancy was attributed to lack of electrical contact by one of the EOG sensors and the corresponding data for the participant's entire session was discarded.

Data analysis further included an analysis of variance (ANOVA) to understand whether there was a statistical effect of Task, Frequency, or trial on the RMS of the EOG signal. The ANOVAs employed the Residual Maximum Likelihood method to adjust for missing values. The model treated the participants as a random effect.

Further, during analysis the stability of the HMD with respect to the participant's head was questioned. To understand helmet motion, a program was created in MATLAB which analyzed the video from the camera to correlate an image of the participant's eye without vibration to images of the same participant's eye that was collected on the same target while undergoing vibration. Vertical offset corresponding to the peak correlation was recorded. This measure provided a sinusoidal output signal which correlated with the amplitude of helmet motion. RMS values from the vertical offset data were calculated and compared to the EOG RMS values.

3.2.4 Results.

The analysis indicated a significant effect of frequency ($F_{(4, 21)}$=26.20; $p<0.0001$) and the Frequency by Task interaction was additionally statistically significant ($F_{(4, 21)}$=6.14; $p=0.002$). Task, approached, but was not statistically significant ($F_{(1, 5)}$=5.84; $p=0.058$). The effect of the Frequency by Task interaction is depicted in Figure 1. As the reader can see, the RMS EOG signal in normalized mV increases as the vibration frequency increases from 2 to 4 Hz and then decreases as the frequency is further increased to 10 Hz. A Tukey post hoc test indicated that the RMS value was statistically higher for the 4 Hz frequency than for any other frequencies. Further, the RMS values for the 6, 8, and 2 Hz conditions were significantly larger than the RMS values for the 10 Hz condition. Finally, the RMS values for the 8, 2, and 10 Hz conditions were lower than for the 6 Hz condition.

The RMS values for each Task at each frequency was compared with a series of ANOVAs. These analyses indicated that the RMS value for Task B was higher than the

37

RMS value for Task A only at the 4 Hz condition ($F_{(1, 5)}=6.71$; $p=0.0464$). The condition at 6 Hz was close to, but not statistically significant ($F_{(1, 5)}=4.49$; $p=0.0881$). Therefore, the variation in eye movement as a function of frequency is greater for the smooth pursuit condition than for the fixation condition at the worst-case conditions. The EOG RMS values are shown in Figure 3.4.

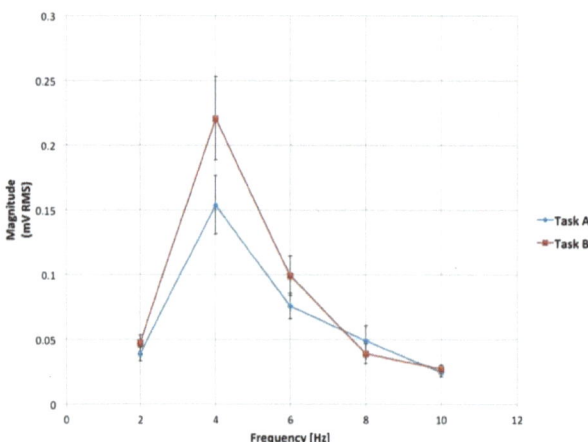

Figure 3.4. EOG RMS Values from Task A and Task B at the corresponding frequencies.

It was necessary to determine if the changes in eye movement magnitudes resulted from head movements or changes in VOR response. Data from the z-axis segment of the accelerometer data for each participant was compared at each vibration frequency. Since the chair vibration input was constant for all tasks, the RMS values in the z-axis for head vibration should be nearly identical. An ANOVA indicated no statistical difference between the head vibration amplitudes as a function of task or Frequency by Task interaction.

The means of the RMS acceleration values are plotted in Figure 3.5 for Tasks A and B. Interestingly, the highest RMS values in vertical head acceleration occurred at 4 and 6

Hz. However, the peak in vertical head acceleration occurred at 6 Hz instead of 4 Hz as found from the eye motion data.

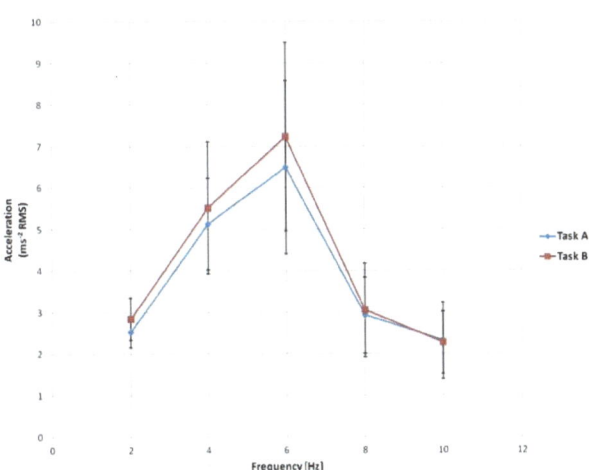

Figure 3.5. Accelerometer RMS Values from Task A and Task B at the corresponding frequencies.

However, analysis of the video recording demonstrated that the helmet moved with respect to the head, referred to as helmet slippage, meaning the display was not stationary on the eye. As the accelerometer data only measured head movement, the amount of helmet motion could only be determined through the video. The vertical offset between an image of the participant's eye while not being vibrated and the image while undergoing vibration was recorded for each recording. A comparison between the EOG RMS values, and the RMS values of the video offset is displayed in Figure 3.6.

Based on the behavior of the EOG RMS values and the video offset RMS values there appears to be a relationship between eye movements and helmet slippage while under vibration. However, more accurate measurements of the helmet slippage and head movements are necessary to better understand this relationship.

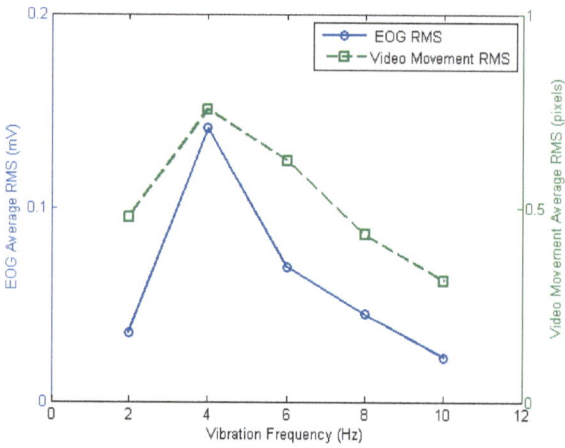

Figure 3.6. EOG RMS values comparison to Video offset RMS values.

3.2.4.1   Initial Discussion.

Previous research to understand the impact of the VOR on visual performance in vibration conditions had relied only on performance measurements, such as reading or aiming errors. This research indicated an increase in aiming error and acuity decrements (30), as well as, a rise in reading errors (12) within the 4 to 6 Hz range. Additionally, the literature suggested that as the vibration frequency increased beyond 6 to 8 Hz, the body would be more successful at dampening vibration and would decrease the transfer of vibration to the head (18).

The results of this study confirmed that the 4-6 Hz vibration conditions are the most destructive to a participant's ability to maintain appropriate fixation on a stationary target or a target engaged with a smooth pursuit eye movement. However, previous research had not distinguished between specific eye tasks, such as a fixating a stationary target versus tracking a moving target. Of importance was the finding that the VOR effect resulted in larger amplitude eye movements when fixating on a target viewed with smooth pursuit eye movements as opposed to fixating on a stationary target. This finding indicates that adding

40

eye motion to the task can have a significant impact on the magnitude of eye motion in this 4-6 Hz range. The measured z-axis, or vertical, head amplitude values confirmed that the vibration in the head was not significantly different between the two tasks. Therefore, this difference in vertical eye movement amplitude cannot be attributed to differences in head motion during the two tasks.

The data indicated that the maximum RMS values for head amplitude, as well as eye movement amplitude occurred in the 4-6 Hz range. However, the peak head acceleration occurred at 6 Hz, which differed from the eye acceleration, which peaked at 4 Hz. This observation verified earlier research findings, which indicated that the eye, head and helmet all vibrate at different frequencies, compounding the difficulty in creating accurate compensation algorithms (21). This study indicated that further research is needed to determine if a correlation exists between the head amplitude and eye movement data. This research needs to include the data of a helmet-mounted accelerometer to verify the differences in vibration.

3.3   Experiment 2

The first experiment demonstrated that EOG values could be related to the frequency input. Analysis of the video recording of the eye from the experiment indicated that the helmet moved with respect to the head while under vibration, indicating helmet slippage should be analyzed in addition to head movement, as it might affect eye movements.

Previous research indicates eye movements do not have a linear relationship with the acceleration level(23). For this reason, it was decided that the follow-on experiment should include multiple acceleration levels. Specifically two different acceleration levels were employed, including 0.1 and 0.2 g. To determine helmet slippage and head movement, each participant wore a helmet-mounted display which was equipped with a set of accelerometers. Additionally, the participants were fit with a bite bar containing a different set of accelerometers and EOG leads were applied around the participant's eyes to permit recording of eye movement data.

3.3.1 Experimental Methodology.

3.3.1.1 Participants.

Twelve participants between 22-29 years old volunteered to take part in the experiment, including 11 male participants. The same qualifications for inner ear, eye, and musculoskeletal system injuries as the first experiment were upheld. One participant experienced an injury, unrelated to the current study between test sessions, and was not permitted to continue with the experiment.

3.3.1.2 Vibration Table.

During the second experiment, participants were again seated on the same single-axis servohydraulic vibration table used during the first experiment. The facility technician ran the vibration table during the experiment, inputting the proper frequencies and displacement values necessary to attain the specified acceleration levels.

3.3.1.3 Helmet.

The display designed for the first experiment was able to display the visual tasks to the participants, but issues with clarity occurred with several participants, indicated the need for a more adjustable display to accommodate a larger variety of participants.

A custom visor was created to fit into the mounts on the helmet, shown in Figures 3.7 and 3.8. The visor's main components were two adjustable side brackets and a plate across the front. The plate attached to a binocular LCD display system that was acquired from a pair of Vuzix Wrap 920 augmented reality glasses and provided 640 by 480 pixels over approximately 31 degree diagonal field of view. The display system had been used in the first experiment to display simple visual tasks to users, simulating the types of task accomplished on a HMD. The plate also had a secondary purpose of blocking a majority of the ambient light, allowing the user to more easily view the display system. The adjustable sides allowed the user to adjust the distance to the display to improve the visibility of the display.

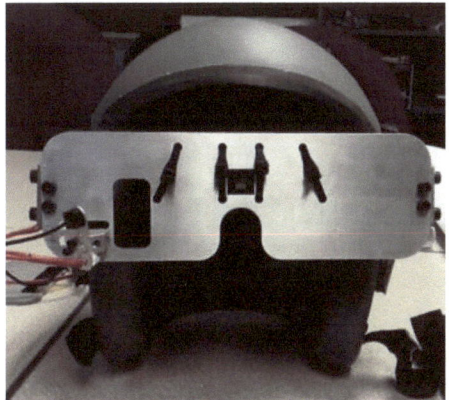

Figure 3.7. Modified helmet from the front

The same EOG procedure was followed as the first experiment since it was proved to be able to provide a consistent measurement of eye movement.

3.3.1.4 Accelerometers.

A triaxial accelerometer pack was used to measure the input acceleration at the floor in the fore-and-aft (X), lateral (Y), and vertical (Z) directions. The pack consisted of three-orthogonally-arranged miniature accelerometers (Entran EGAX-25) embedded into a two-piece plastic disk. Two sets of six miniature accelerometers (Entran EGA 125-10D)

Figure 3.8. Modified helmet from the side.

were orthogonally-arranged and strategically glued inside of the two-piece plastic mount, and attached to a bite bar and the top of the helmet. These accelerator packs permitted the measurement of the translational and rotational movements of the user's head and helmet. The bite bar used a custom impression of each participant's teeth, obtained using molding clay and a mouthpiece for a tight fit. This custom fit bite bar permitted accurate accelerometer readings of the participants head motions. The slippage between the head and helmet motions were calculated by differencing values obtained from the helmet-mounted and bite bar-mounted accelerometer packs. These slippage values indicate how much the display moved with respect to the participants' head.

All acceleration data were collected onto a 48 channel data acquisition system (EME, Corp), filtered at 250 Hz, and digitized at 1024 samples per second. The EOG and accelerometer data were collected simultaneously, triggered for the 15 seconds of task completion at each frequency.

3.3.2 Experimental Procedure.

Upon arriving, participants completed an informed consent and underwent appropriate screening procedures. The participants were then seated on the acceleration table and instrumented as necessary. The instrumentation was then calibrated to insure the proper acceleration level at each frequency. The participants then trained and completed the first of two test sessions.

During each test session, the participants first completed an eye movement calibration without vibration. The calibration required the participants to fixate on a target at a sequence of five locations, indicating the center and the extremes of target location during the experiment. The measurements from the calibration period were used to understand the magnitude of the EOG signals for each participant. The same format for the tasks was used as the first experiment, but a dot was used as the fixation target to avoid the potential for multiple fixation points from the first experiment. Upon completing the tasks at one acceleration level, the instrumentation was removed from the participants; they were debriefed and permitted to leave. Participants then completed the second test session on a different day, following the same procedure, only experiencing a different acceleration level.

The participants were exposed to six different sinusoidal vibration frequencies, including 0, 2, 4, 6, 8, and 10 Hz. These frequencies were experienced at 0.1 g ($\sim$0.69 m/s^2 rms) peak acceleration level for one test session and 0.2 g ($\sim$1.37 m/s^2 rms) peak acceleration level for the other test session. The order the acceleration level were experienced by the participants was counterbalanced to avoid any biasing effects. These vibration levels and their durations were selected in accordance with the current human exposure guidelines (14), with overall levels indicating a low risk of health effects. The procedure was reviewed and approved through the Air Force Institutional Review Board process.

3.3.3 Data Analysis.

For data analysis of the tracking task, only the sections of the signal where the dot was moving horizontally were considered. During horizontal tracking, any vertical eye movements could be attributed to compensating for the vibration. If one of the horizontal segments had a non-representative eye movement, such as a blink, in the signal, that segment was not used in calculation. For the fixation task, blinks were segmented out of the signal, and the remaining signal was used for analysis. A high-pass filter was created in MATLAB using the filterbuilder function. The stop frequency was 1 Hz, and the pass frequency was 1.5 Hz, in order to eliminate the effects of drift in the signal, but retain any effects from the applied frequencies. Then, the signal was normalized using a value calculated from the calibration task. The normalization value was half the difference between the 90th percentile value of the EOG signal when the dot was at the top of the screen and the 10th percentile value when the dot was at the bottom of the screen. The normalization value estimated the extremes of the eye movement in the EOG signal.

Head and helmet rotational accelerations were calculated using the difference between the time histories of the two orthogonally placed accelerometers of interest, and dividing by the moment arm. The rotational displacements of the head and helmet were estimated using the Fast-Fourier Transform (FFT) of the time history of the accelerometer signal, dividing by the square of the angular frequency, and multiplying by -1. The inverse FFT was applied to obtain the estimation of the displacement. The helmet slippage rotational accelerations and displacements were calculated as the difference between the head and helmet time histories.

PSD analyses using Welch's method were conducted on the filtered segments of the EOG signal and the accelerometer values to compare the applied vibration frequency to response in the frequency domain. The magnitude of the power of the PSD was converted to RMS values at each frequency by multiplying the PSD values by the resolution and taking the square root. The RMS value at the applied frequency was obtained, and was used in further analysis to compare the head and helmet motions to the eye movements. The EOG RMS values for one subject were more than 3 standard deviations from the mean, and were not included in the analysis.

A repeated measures ANOVA was conducted on the EOG RMS values and the associated accelerometer RMS values for vertical head acceleration, head and helmet slippage pitch accelerations, and head and helmet slippage pitch angles. The LSD comparison on within measures was completed to determine the significant difference between the levels. When Mauchly's test for sphericity failed, the Greenhouse-Geisser correction was applied to the analysis. A significance level of 0.05 was used for this analysis.

3.3.4 Results.

The effect of frequency ($F(2.097, 18.875)=11.64$; $p \leq 0.000$) was significant for the EOG RMS values. The effect of task ($F(1, 9)=0.385$; $p=0.550$), acceleration level ($F(1, 9)=2.182$; $p=0.174$), and the interaction terms on the EOG RMS values were not significant. The average EOG RMS values for the fixation task, Task A, and the tracking task, Task B, are shown in Figure 3.9. A pairwise comparison between the frequencies shows the EOG RMS values at 4 Hz are significantly larger than the values at 2, 8 and 10 Hz. The EOG RMS values at 4 and 6 Hz are approaching significantly different, with a p-value of 0.065.

Comparing task, frequency and acceleration level of the vertical head acceleration RMS values using repeated measures analysis indicated that the effects of frequency ($F(2.195, 19.758)=34.774$; $p \leq 0.000$), acceleration level ($F(1, 9)=380.664$; $p \leq 0.000$), and the interaction between frequency and acceleration level ($F(4, 36)=15.003$; $p \leq 0.000$) had a significant effect on the RMS values. The effects of task ($F(1, 9)=1.325$; $p=0.279$), and the additional interactions were not significant. The vertical head acceleration RMS values are

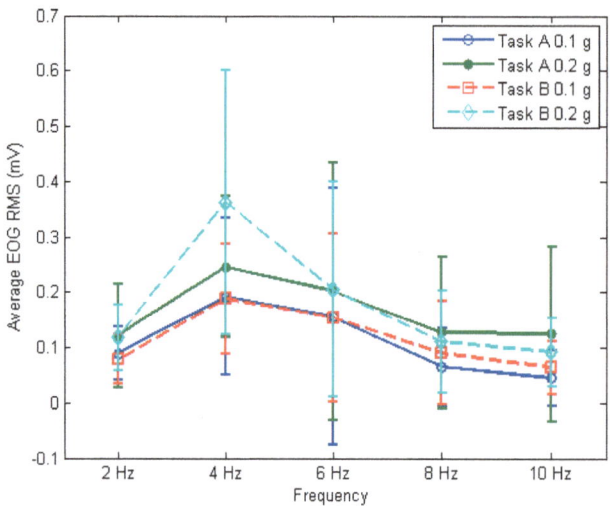

Figure 3.9. Average EOG RMS values for each task and each acceleration level at the corresponding frequency.

shown in Figure 3.10. ANOVAs on the change in frequency, separated by acceleration level determined that for 0.1 g, 6 Hz was significantly larger than any other frequency. For 0.2 g, 4 Hz was significantly larger than all other frequencies. Six Hz was significantly different from all other frequencies.

The effects of frequency ($F_{(1.476, 13.284)}=14.139$; $p=0.001$) and acceleration level ($F_{(1, 9)}=5.106$; $p=0.050$) are significant for the head pitch acceleration RMS values. The effects of task ($F_{(1, 9)}=1.513$; $p=0.250$) and all interaction terms were not significant. The head pitch acceleration RMS at 2 Hz was significantly less than for all other frequencies, and 4 and 6 Hz were greater than all other frequencies. The head pitch acceleration RMS values are shown in Figure 3.11.

The helmet slippage pitch acceleration RMS values are shown in Figure 3.12, the effect of frequency is significant ($F_{(1.739, 15.652)}=6.675$; $p=0.010$), while, the effects of task

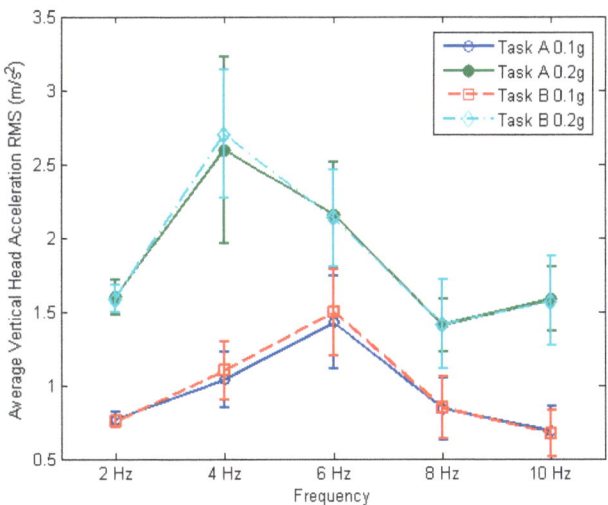

Figure 3.10. Average vertical head acceleration RMS values for each task and each acceleration level at the corresponding frequency.

(F(1, 9)=0.279; p=0.610), acceleration level (F(1, 9)=2.637; p=0.139) and all interaction terms were not significant. For both tasks and acceleration levels, 2 Hz is significantly less than all other frequencies.

For the head pitch angle RMS values, frequency (F(1.746, 15.718)=53.429; p≤0.000), acceleration level (F(1, 9)=35.261; p¡=0.000), task and frequency interaction (F(2.348, 21.131)=4.473; p=0.020), and the interaction between frequency and acceleration (F(2.240, 20.162)=9.902; p=0.001) are significant. Task (F(1, 9)=0.000; p=0.989) and the remaining interactions are not significant. Separating the RMS values by task, for the fixation task, frequency (F(1.88, 16.89)=45.86; p≤0.000), acceleration (F(1, 9)=36.70; p≤0.000) and the interaction (F(2.39, 21.55)=7.92; p=0.002) are significant. ANOVAs on each acceleration level indicate for both 0.1 g and 0.2 g, the RMS values at 2 and 4 Hz are significantly larger than the RMS values at the other frequencies. For the tracking task, frequency (F(1.759, 15.831)=45.266; p≤0.000), acceleration level (F(1, 9)=24.833; p=0.001), and

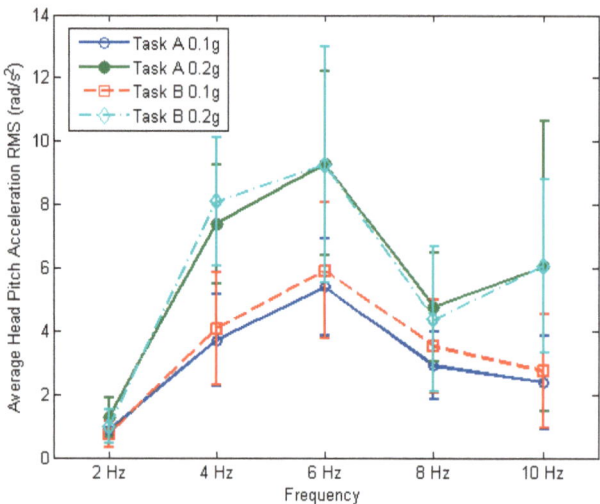

Figure 3.11. **Average head pitch acceleration RMS values for each task and each acceleration level at the corresponding frequency.**

the interaction term (F(1.869, 16.819)=8.159; p=0.004) are all significant. At both 0.1 g and 0.2 g, 8 and 10 Hz are significantly less than the other frequencies. At 0.2 g, 4 Hz is significantly larger than all other frequencies. Similar results were found from the ANOVAs conducted on the interaction of frequency and acceleration level. The head pitch angle RMS values are shown in Figure 3.13.

Frequency (F(1.804, 16.240)=26.744; p≤0.000) and acceleration level (1, 9)=17.246; p=0.002) are significant for the helmet slippage pitch angle RMS values. Task (F(1, 9)=0.152; p=0.706) and the interactions are not significant. A pairwise comparison demonstrates that the RMS helmet slippage pitch angle at 2 Hz is significantly larger than all other frequencies. The helmet slippage pitch angle RMS values are shown in Figure 3.14.

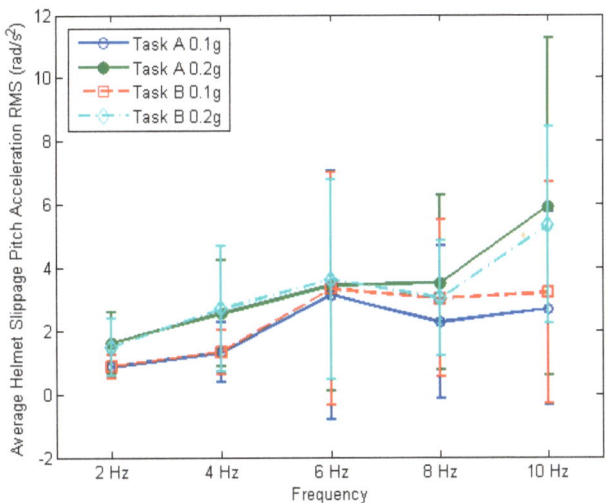

Figure 3.12. Average helmet slippage pitch acceleration RMS values for each task and each acceleration level at the corresponding frequency.

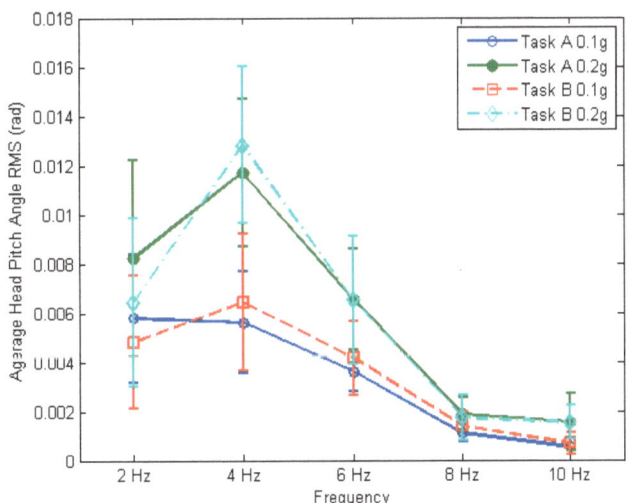

Figure 3.13. Average head pitch angle RMS values for each task and each acceleration level at the corresponding frequency.

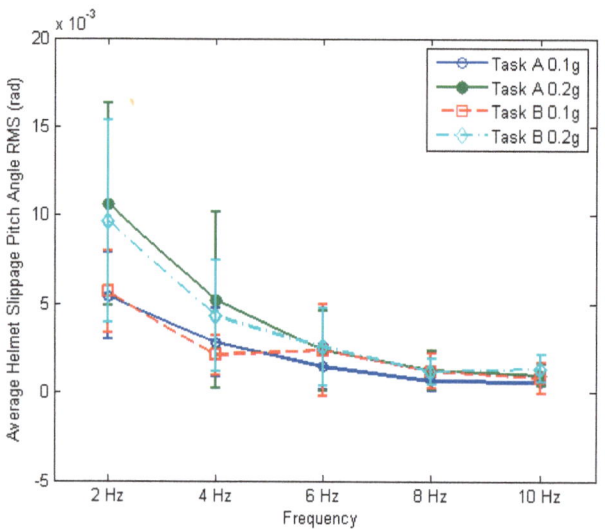

Figure 3.14. Average helmet slippage pitch angle RMS values for each task and each acceleration level at the corresponding frequency.

3.4 Discussion

As the accelerometer values between the two visual tasks are similar for all conditions, it is expected that the vestibular system received approximately the same signals, and the VOR should respond in a similar fashion, regardless of task as was observed in this second experiment. The change in the EOG RMS values appears higher at the higher acceleration level, however the acceleration level is not significantly different. As only the change in frequency had an effect on the EOG RMS values, while all of the accelerometer values, except helmet slippage pitch acceleration, are significantly affected by the other factors, this indicates that the eye is not as affected by the changes in vibration amplitude or that the noise level is larger for the high acceleration levels.

The variables that have the strongest relationship with the EOG RMS values were determined using a stepwise regression. The frequency, acceleration level, and accelerometer values were included in the analysis. Head pitch acceleration and helmet slippage pitch acceleration were determined to be the significant factors to predict the EOG RMS values for the fixation task. The estimated regression function has an adjusted R^2 of 0.234 and is:

$$FixationEOG = 0.0973 + 0.0236 \cdot HeadPitchAcceleration - 0.0235 \cdot HelmetSlippagePitchAcceleration$$

$$(3.1)$$

Using these two input variables to create a regression function to estimate the EOG measurements from the tracking task results in the function:

$$TrackingEOG = 0.0909 + 0.0260 \cdot HeadPitchAcceleration - 0.0229 \cdot HelmetSlippagePitchAcceleration$$

$$(3.2)$$

This function has an adjusted R^2 of 0.285. As there was not a significant effect from task, it was expected that the estimated regression functions would be similar. Although Equations 3.1 and 3.2 are not able to be used to directly compute eye movements, they are useful to understand the factors that have a relationship with the eye movements. Understanding what vibration induced head and helmet movements cause eye movements is a crucial aspect in this area of research, and knowing the relationship between head pitch acceleration, helmet slippage pitch acceleration, and eye movements could greatly benefit future studies.

The EOG RMS values for Experiments 1 and 2 peaked for both visual tasks at 4 Hz. Since Experiment 1 used a target acceleration level of 0.1 g, the 0.2 g RMS values from Experiment 2 are not comparable. At 0.1 g for both tasks, the EOG RMS values at 4 and 6 Hz are not significantly different, but with greater number of participants, more significant results could occur. The EOG signal contains a large amount of noise that results in large variabilities in EOG RMS values. The RMS values from both experiments support the documented blur experienced in HMDs while under vibration, especially in the 4-6 Hz range (18).

The accelerometer RMS values between Experiments 1 and 2 demonstrate similar results for both tasks at the 0.1 g acceleration level, peaking at 4 Hz.

3.5 Conclusion

There is not a significant difference between the eye movements depending on the visual task being completed. This does not support the results of the research applying filters to counteract the vibration, which found that a single filter was not capable of effectively stabilizing the image displayed for different visual tasks. However the peak EOG RMS values at 4 and 6 Hz support the reports of low frequency vibrations affecting visual performance, especially in the 4-6 Hz range. As the difference between the EOG RMS values at 4 and 6 Hz approaches significance, the relationship between them could be further investigated with greater power by increasing the number of participants.

The difference in the vertical acceleration RMS values as the acceleration level changes demonstrates a non-linear relationship between body resonance and the vibration level applied, supporting the work by Smith, Mosher, et al. (23). As only two acceleration levels were tested, further research would be able to compare how body resonance is affected by acceleration level changes.

The similar results for the estimated regression functions for the EOG RMS values for the fixation task and the tracking task demonstrates that there is a relationship between the head pitch acceleration, the helmet slippage pitch acceleration and the EOG RMS values. Although EOG provides a consistent method to measure eye movements during vibration, the signal includes subject-specific noise and apparent drift, making it difficult to determine accurate eye location. Accurate video-based eye tracking demands that the camera remains relatively stationary with respect to the head, or at least avoid rapid changes in head location. The microcamera attached to the helmet suffered from the helmet slippage movements which affected the location of the eye on the video, as well as the changes in head shape between the participants which affected how clear an image of the eye was recorded. A more robust design for the camera, a method to eliminate helmet slippage, or a more effective method to analyzing EOG signals is necessary to determine absolute eye position.

IV. Conclusion

4.1 Chapter Overview

This chapter provides an overall summary of the research and results covered in this thesis. It recaps the significant results from a previous study to a human subject experiment that further expanded research into human head, eye, and helmet motion in the presence of vibration. Additionally, the military and commercial importance of the research and results from the experiment are described. Finally, suggestions on future work that would further knowledge in this area of research are outlined and justified.

4.2 Summary of Results

Analyzing the video data recorded from previous research and comparing these results to the VOR effects at different frequencies indicated that the video, and therefore the display, was moving with respect to the participant's head. This was caused by the movement of the helmet, which is allowed for sizing and comfort issues. Comparing each video frame to a standard image taken from a non-vibration period provides an estimate of how much the camera moved while under vibration. Comparing the RMS values of the video to the EOG RMS values indicated that a relationship between the helmet motion and the eye movements could possibly exist. As a result, a human-subjects experiment was designed to probe this relationship as well as to extend the experimental conditions to include multiple accelerations.

The effectiveness of using EOG values to analyze eye movements was found to be difficult since the EOG signal contains substantial noise and drift. In addition, EOG signals are characteristic for each individual, making it difficult to compare signals between subjects. Overall, EOG signals are able to indicate general movements, but do not always indicate accurate eye position. By applying a 0.5 second moving average filter, the low frequency drift was removed without affecting the movements associated with the vibration input. Normalizing the data attempted to scale the eye movements so the signal could be compared between subjects.

Analysis of the EOG RMS values in this study demonstrated that only frequency affects eye movements. The fact that task did not change the accelerometer values indicates that the inner ear canals received the same signals, which would induce the same VOR response. Any change in the eye movements from the fixation task to the tracking task can possibly be attributed to either the suppression of the VOR, or to the combination of the pursuit response and the VOR movements acting in coordination with each other. The EOG RMS values for both tasks experienced peak values at 4 Hz, which was not significantly different from 6 Hz. These results coincide with the reported degradation in visual performance during vibrations between 4 and 6 Hz (18).

The vertical head acceleration RMS values describe how the head moves up and down while undergoing vibration. The similarity of the values across tasks indicates that the body responds the same when the same vibration conditions are applied. However, the significant difference between the RMS values at different acceleration levels demonstrates that the body's response changes. At 0.1 g, the peak vertical head acceleration occurs at 6 Hz, while at 0.2 g, the peak acceleration occurs at 4 Hz. These values could help to explain the conflicting results from previous research that attempt to describe the body's response at different vibration levels (22; 12). The body's response appears to change as a function of acceleration level input, meaning previous research that used different vibration levels would obtain differing results.

Using a stepwise regression analysis, the head pitch acceleration RMS and helmet slippage pitch acceleration RMS values were determined to have a significant relationship with the EOG RMS values for the fixation task. The relationship between the head pitch acceleration values and the EOG values was expected based on the VOR response to rotational motions perceived in the inner ear canals. The helmet slippage pitch angle RMS values were expected to have a more significant relationship than was shown in past literature based on the results of the video analysis from the previous study. This analysis confirms that the blur that is present in HMDs in vibratory environments likely stems from a combination of vestibularly-induced eye movements and the movement of the helmet-mounted display.

4.3   Implications

Understanding how the eyes move at the different vibration conditions despite the task being completed is the most important focus in this area of research currently. As indicated by the similar estimated regression functions developed, there are similarities in how the eye moves regardless of task. Validating the recorded experiences of HMD users that experience the largest vision decrements in the 4-6 Hz regime is an important result of this research. Although the regime was noted, the reasoning behind the vision decrements could only be theorized, but by understanding that the eye movements were largest at those frequencies, a more complete profile of the eye's response can be used to compensate for vibrations. Although the results that eye movements increase when the acceleration level increases were expected based on the larger displacement values necessary to reach the higher acceleration level, the ability to determine how much more the eyes moved at the larger acceleration level is essential for compensation algorithms.

In spite of the low adjusted R^2 values for the estimated regression functions, knowing which values indicate eye movement focuses future research. Simply the development of a function that is able to describe eye movement while under vibration is a positive

58

conclusion. Although more research into accurately decomposing EOG signals into eye movements is necessary to be able to use the time history of the signal, the estimated regression functions provide a starting point for understanding the relationship between the head, helmet, and eye movements.

4.4   Recommendations for Future Work

The new design for the display attachment placed the display on the inside of the attachment, unlike the previous design. While this allowed for participants to place the display in a position more similar to the original purpose of the display, it also limited the visibility of the camera intended to record the eye throughout the experiment. Various adjustments might improve the overall design of the helmet attachment. Compromising between the original design and the one used for this research, pulling the display away from the eyes allows for improved angles for the camera may permit higher quality video of the eye. This could potentially lead towards the ability to use the video for eye tracking during the tasks. Another improvement is the addition of rotation of the display in and out of participant's line of sight. Currently the rotation about the attachments to the helmet and the adjustable sides of the attachment add flexibility in the display to conform to the individual fit of each participant's face, but additional flexibility would improve the participants' ability to see the stimuli displayed, resulting in more accurate eye movement measurements. In addition, developing a method to minimize the amount of helmet slippage, such as an oxygen mask, would increase the understanding of the relationship between eye movements and helmet slippage.

The fixation task demonstrates a simple version of the type of task HMD users commonly complete. However, in a realistic environment, information is displayed on the entire display. The change in eye movements by fixating on off-center positions would contribute to understanding the compensation methods necessary to counteract vibration effects.

Previous research recorded that larger head motions occurred as a result of vibration in off-axis head orientations. These larger head motions also resulted in large helmet slippage values (25). Based on the results of this research, alternate head positions should result in larger eye movements. As HMD users are able to move their head in all directions and the information displayed will still be available, analysis of how head orientation affects eye movements while under vibration is crucial to formulating a complete picture of eye movements to compensate for them.

As discussed earlier, the EOG signals are not easily analyzed and compared between participants. Each participant's EOG signal has individual features, noise levels, and clarity of characteristics such as blinks, all of which results in large variances in the RMS values. Understanding the reasons behind the differences in the EOG signals between the subjects could improve data collection in the future. Further research is necessary to more effectively analyze EOG signals so subjects can be compared, however basic analysis is possible currently.

4.5   Final Thoughts

The research conducted delves into a crucial area for future HMD use, analyzing the components that must be considered to effectively compensate for vibrations. The results help support various other findings, while also expanding into novel areas that have not been covered previously. This research demonstrated that eye movements peak in the expected 4-6 Hz range, as well as the effect acceleration level and task have on the eye movements. The military and commercial applications of HMDs are expanding, and quality of imaging is a determining factor of the use of certain technologies over others. The ability

60

to present legible information regardless of vibration conditions would increase the amount of information able to be displayed, potentially increasing the user's performance.

Bibliography

[1] "Joint Strike Fighter (JSF) HMD System", May 2009. URL http://www.jsf.mil.

[2] "Consumer Products", 2014. URL http://www.lumus-optical.com/index.php?option=com_content&task=view&id=9&Itemid=15.

[3] "Glass", 2014. URL http://www.google.com/glass/start/how-it-looks/.

[4] "STAR 1200XLD", 2014. URL http://www.vuzix.com/augmented-reality/products_star1200xld/.

[5] Adelstein, Bernard D, Mary K Kaiser, Brent R Beutter, Robert S McCann, and Mark R Anderson. "Display strobing: An effective countermeasure against visual blur from whole-body vibration". *Acta Astronautica*, 2012.

[6] Babcock, Jason S and Jeff B Pelz. "Building a lightweight eyetracking headgear". *Proceedings of the 2004 symposium on Eye tracking research & applications*, 109–114. ACM, 2004.

[7] Bayer, Michael M, Clarence E Rash, and James H Brindle. "Introduction to helmet mounted displays". *Helmet-mounted displays: Sensation, perception, and cognitive issues*, 47–108, 2009.

[8] Daetz, Daniel. *Development of a biodynamic interference suppression algorithm for a helmet-mounted display tracking task in the presence of aircraft buffet*. Master's thesis, Air Force Insistute of Technology, Wright-Patterson, AFB, OH, 2000.

[9] Diez, Pablo F, Eric Laciar, Vicente Mut, Enrique Avila, and Abel Torres. "A comparative study of the performance of different spectral estimation methods for classification of mental tasks". *Engineering in Medicine and Biology Society, 2008. EMBS 2008. 30th Annual International Conference of the IEEE*, 1155–1158. 2008.

[10] Geiselman, Eric E and Paul R Havig. "Rise of the HMD: the need to review our human factors guidelines". *SPIE Defense, Security, and Sensing*, 804102–804102. International Society for Optics and Photonics, 2011.

[11] Griffin, J., R.W. McLeod, J. Moseley, and C.H. Lewis. *Whole-body vibration and aircrew performance*. Human Factors Research Unit, Institute of Sound and Vibration Research, University of Southampton, 1986.

[12] Griffin, Michael J. *Handbook of Human Vibration*. Academic press, 1990.

[13] Griffin, M.J. and C.H. Lewis. "A review of the effects of vibration on visual acuity and continuous manual control, part I: Visual acuity". *Journal of sound and vibration*, 56(3):383–413, 1978.

[14] International Organization for Standardization (ISO). *Mechanical vibration and shock - Evaluation of human exposure to whole-body vibration - Part 1: General requirements. ISO 2631-1:1997*, 1997.

[15] Matthes, R, M Feychting, A Ahlbom, E Breitbart, R Croft, FR de Gruijl, A Green, M Hietanen, K Jokela, JC Lin, et al. "Icnirp guidelines on limits of exposure to incoherent visible and infrared radiation". *Health Physics*, 105(1):74–91, 2013.

[16] Moseley, M and M. Griffin. "Whole-body vibration and visual performance: An examination of spatial filtering and time-dependency". *Ergonomics*, 613–626, 1987.

[17] Parr, Jeffrey C, Michael E Miller, Joseph A Pellettiere, and Roger A Erich. "Neck Injury Criteria Formulation and Injury Risk Curves for the Ejection Environment: A Pilot Study". *Aviation, space, and environmental medicine*, 84(12):1240–1248, 2013.

[18] Rash, Clarence, Keith Hiatt, Robert Wildsunas, J. Lynn Caldwell, Melvyn Kalich, Gregory Lang, Ronal King, and Robert Noback. "Perceptual and Cognitive Effects Due to Operational Factors". *Helmet-Mounted Displays: Sensation, Perception and Cognition Issues*. U.S. Army Research Laboratoy, Fort Rucker, AL, 2009.

[19] Rash, Clarence, William McLean, John Mora, and Melissa Ledford. *Design Issues for Helmet Mounted Display Systems for Rotary-Wing Aviation*. Technical report, U.S. Army Aeromedical Research Laboratory, Fort Rucker, AL, 1998.

[20] Roy, Jefferson E and Kathleen E Cullen. "A neural correlate for vestibulo-ocular reflex suppression during voluntary eye–head gaze shifts". *Nature neuroscience*, 1(5):404–410, 1998.

[21] Seagull, F.J and C.D. Wickens. "Vibration in command and control vehicles: visual performance, manual performance, and motion sickness: A review of the literature". *Human Factors Division, Institute of Aviation, University of Illinois*, 2006.

[22] Shoenberger, Richard. "Human Response to Whole-Body Vibration". *Perceptual and Motor Skills*, 34:127–160, 1972.

[23] Smith, S. D., S. E. Mosher, D. R. Bowden, A.Y. Walker, and J. G. Jurcsisn. "Head Transmissibility Characteristics During Single- and Combined-Axis Vibration Exposures". *Proceedings of the 42nd United Kingdom Conference on Human Responses to Vibration*, 2007.

[24] Smith, Suzanne. "Collection and Characterization of Pilot and Cockpit Buffet Vibration in the F-15 Aircraft". *SAFE Journal*, 30(3):202–218, 2002.

[25] Smith, Suzanne and Jeanne Smith. "Head and Helmet Biodynamics and Tracking Performance in Vibration Environments". *Aviation, Space, and Environmental Medicine*, 77:388–397, 2006.

[26] Stuart, Geoffrey W, Ken I McAnally, and James W Meehan. "Head-up displays and visual attention: integrating data and theory". *Contemporary Issues In Human Factors And Aviation Safety*, 25, 2005.

[27] Uribe, Daniel. *An investigation and analysis of the vestibulo-ocular reflex (VOR) in a vibration environment.* Master's thesis, Air Force Insistute of Technology, Wright-Patterson, AFB, OH, 2013.

[28] Uribe, Daniel J and Michael E Miller. "Eye Movements When Viewing a HMD Under Vibration". *Proceedings of the Human Factors and Ergonomics Society Annual Meeting*, volume 57, 1139–1143. SAGE Publications, 2013.

[29] Uribe, Daniel J, Michael E Miller, and Suzanne Smith. "An analysis of the vestibule-ocular reflex during vibration". *17th International Symposium on Aviation Psychology*, 17. 2013.

[30] Velgar, Mordekhai. *Helmet-Mounted Displays and Sights*. Artech House, 1998.

[31] Vercher, J.L., G.M. Gauthier, E. Marchetti, P. Mandelbrojt, and Y. Ebihara. "Origin of eye movements induced by high frequency rotation of the head." *Aviation, space, and environmental medicine*, 55(11):1046–1050.

[32] Wells, M.J. and M.J. Griffin. "Vibration-induced eye motion". *Aerospace Medical Association Annual Scientific Meeting, Houston, Texas*, 23–26. 1983.

[33] Wong, Jo Yung. *Theory of ground vehicles*. John Wiley and Sons, 2001.

Appendix A: Screening Questionnaire

Screening Checklist

Subject ID:

Age:

Gender:

Neck Measurement (in):

Forehead Measurement (in):

Height:

(Women Only) Are you pregnant or suspect to be pregnant?   YES/NO

(Women Only) Do you have breast implants?   YES/NO

Do you wear corrective lenses (glasses/contact lenses)?   YES/NO

Have you had corrective eye surgery (PRK/LASIK)?   YES/NO

Have you been diagnosed or treated for any eye injuries or disease(s)?   YES/NO

Have you been diagnosed or treated for any inner ear injuries or disease(s)?   YES/NO

Have you experienced any inner ear problems in the past month (vertigo, dizziness, infection)?
YES/NO

Have you consumed alcohol in the past 24 hours?   YES/NO

Are you currently experiencing or in the past month have experienced:

…cold or allergy congestion symptoms?	YES/NO
… pain in the musculoskeletal system especially in the back or neck?	YES/NO
… Numbness/Tingling/Weakness in Extremities?	YES/NO
… Constant Headaches?	YES/NO
… Shooting Pain into Arms/Hands/Legs/Feet?	YES/NO

If any above are YES, explain below:

Date reviewed:
Signature of research monitor:

Appendix B: Informed Consent Document

Informed Consent Document
for
An Investigation and Analysis of the Vestibulo-ocular Reflex (VOR) in a Vibration Environment

AFIT/ENV, Wright-Patterson AFB

Principal Investigator: Michael E. Miller/Ph.D/Assistant Professor, AFIT/ENV, (937) 255-3636 x4651, Michael.Miller@afit.edu

Associate Investigators: Kalyn Tung/2LT/Master's Student, AFIT/ENV, (937) 255-3636 x4651, Kalyn.Tung@afit.edu

Suzanne D. Smith, PhD, Senior Biomedical Engineer, AFRL/RHCP, DSN 785-9331, Suzanne.smith@wpafb.af.mil

Research Monitor: Capt Dawn M. Russell, USAF, MPAS, PA-C, AFMC, 711 HPW/RHCP, DSN 798-3724 dawn.russell@wpafb.af.mil

Medical Observers: TSgt Bethany L. Repp, USAF, AFMC, 711 HPW/RHCP, DSN 798-3702 Bethany.repp@wpafb.af.mil

SSgt Andrew J. Jimenez, USAF, AFMC, 711 HPW/RHCP, DSN 798-3147 Andrew.jimenez@wpafb.af.mil

SSgt Misty A. Hobbs, USAF, AFMC, 711 HPW/RHCP, DSN 798-3146 misty. hobbs@wpafb.af.mil

1. **Nature and purpose:** You have been offered the opportunity to participate in the research study entitled "An Investigation and Analysis of the Vestibulo-Ocular Reflex (VOR) in a Vibration Environment" research study. Your participation will occur sometime between 1 Nov 2013 and 31 Dec 2013, at the Single-Axis Servohydraulic Vibration Facility located in Building 824, Area B, Wright-Patterson AFB.

   The purpose of this research is to measure eye movement associated with the VOR during low frequency vibration while performing various targeting tasks using a Helmet Mounted Display (HMD). This research seeks to provide a greater understanding of the effects of vibration on eye movement and visual performance degradation when using HMDs, and to generate baseline data for developing effective compensation techniques to mitigate this degradation.

An Investigation and Analysis of the Vestibulo-ocular Reflex (VOR) in a Vibration Environment
FWR20130014H, Version 3.00
AFRL IRB Approval Valid from 15 Oct 2013 to 14 Oct 2014

The time requirement for each volunteer subject is anticipated to be a total of 2 visits of approximately 40 minutes each. A total of approximately 15-20 subjects will be enrolled in this study.

2. **Experimental procedures:** You will wear ABUs during the test sessions. You will be escorted onto the Single-Axis vibration table, seated in the seating system, and loosely restrained. There will be a rubber pad mounted onto the seat back cushion. The pad has miniature sensors embedded in them for monitoring the vibration that enters your body. A pad is also installed on the seat pan. You will also wear a helmet with a miniature speaker located in the ear cuff for hearing commands from the performance task. The helmet will be mounted with a camera, an infrared light source and a video display that will be used for performance tasks.

The low-level vibration exposure signals will be recreated on the Single-Axis vibration table. The selected signal will vary between 0-10 Hz. This level of vibration is akin to what a subject would encounter in an everyday, real-world scenario, such as traveling in a car over a rough or unpaved road, or when encountering brief, minor turbulence in a commercial aircraft.

Each test session will consist of three different tasks that you will perform one at a time. The tasks include a single-point fixation task, a smooth tracking of a target task, and a tracking of a jumping target task. You will be asked to perform each task, not necessarily in the order described, one after another with a two-minute break between. The tasks are expected to take approximately 3 minutes each for a total of 9 minutes of task performance. You will be instructed on what to do for each task before the session begins and reminded of what the upcoming task will be during each break.

Representation of Tasks

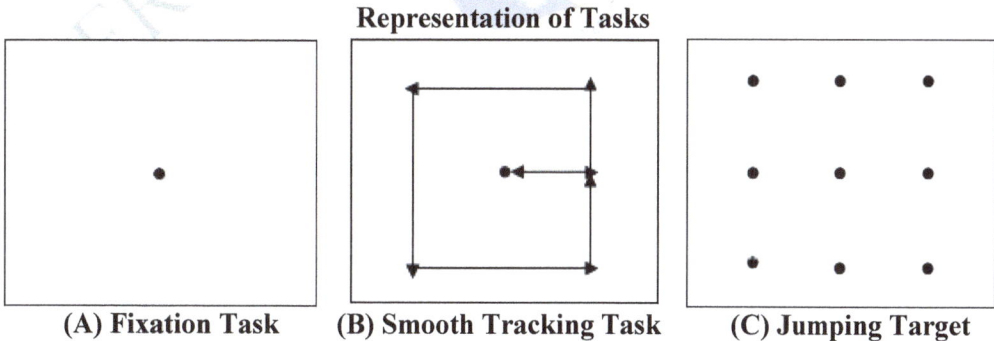

| (A) Fixation Task | (B) Smooth Tracking Task | (C) Jumping Target |

During the formal test session, setup will take approximately 10-15 minutes. When setup is complete you will be instructed of the order of your tasks and reminded what the first task will be and what to do. You will then be exposed to low-level vibrations increasing in specific increments over a determined range or to vibrations varying by a specific increment over a determined range. You will be given a 20 second acclimation period for each interval and then you will be prompted to begin the task and continue the task for an additional 15 seconds at that specific frequency interval. The vibration will then increase or decrease to the

An Investigation and Analysis of the Vestibulo-ocular Reflex (VOR) in a Vibration Environment
FWR20130014H, Version 3.00
AFRL IRB Approval Valid from 15 Oct 2013 to 14 Oct 2014

next interval depending on the session and the same 20 second acclimation period will be given followed by 15 seconds of performing the task. This will occur a total of 6 times including one interval at 0 Hz, or no vibration. You will be given a two-minute break, where you are to remain seated and lightly restrained. You will be reminded on what the next task will be and how to perform it. The process will start over again and will be repeated identically for the three tasks. When all the tasks have been completed, the session will end and you will be asked to complete a post-test questionnaire. You will be observed for any signs of dizziness or other side effects, prior to exiting the facility. There will only be one formal test session per day.

3. **Discomfort and risks:** Due to the very short duration of exposure to the vibration signals the probability is low that you will experience any discomfort. However, if you do, you may have feelings of annoyance and discomfort in the thighs, buttocks, back, and neck, with some muscle aches. You can stop the test at any time by notifying the test conductor. You should discontinue testing if you experience excessive discomfort or dizziness.

4. **Precautions for female subjects or subjects who are or may become pregnant during the course of this study:** If you are a female, you must read this section prior to signing the consent form. There is little information available concerning the effects of vibration on pregnant females. However, female military members in the USAF are exposed to operational environments that include short and prolonged periods of vibration. Therefore, there is a real need to assess the effects of vibration on females for the improvement of exposure standards and equipment design criteria. The following are specific precautions that apply to female subjects:

   a. Pregnancy - There are no data with which to evaluate the risk to a developing fetus (spontaneous abortion or fetal abnormalities) of exposure to vibration. Pregnant females cannot participate in vibration studies, no matter what the level of the exposure. Pregnancy will be determined by pregnancy tests administered and read by a trained medical observer prior to being cleared for testing. It is appropriate to utilize an effective contraceptive technique prior to, and for the duration of, vibration exposures as a human subject. If you become pregnant or feel you might be pregnant, contact your medical provider and the study investigator or research monitor.

   b. Contraceptives - The use of oral contraceptives in the general population has been implicated in an observed increased incidence of medical problems such as inflammation of the large veins in the legs and pelvis with formation of blood clots. These clots may be dislodged and travel to the lungs with a potentially fatal outcome. Current medical studies examine these problems in a normal environment. Medical studies have also suggested that smoking and the use of oral contraceptives place the female subject at a greater risk. No studies have been done to examine the influence, if any, of vibratory motion on the use of oral or intrauterine contraceptives.

   c. Ovarian Abnormalities - The ovary is subject to cystic enlargement and other conditions that may occur with or without symptoms. There is a possibility that prolonged

An Investigation and Analysis of the Vestibulo-ocular Reflex (VOR) in a Vibration Environment
FWR20130014H, Version 3.00
AFRL IRB Approval Valid from 15 Oct 2013 to 14 Oct 2014

exposures to vibration could increase the normal risk that such an enlarged cyst may burst or that the ovary may twist about its support, cutting off the blood supply. This situation would possibly require major surgery to correct the condition with the attendant risks of loss of the involved ovary, bleeding, infection, or death.

    d.   Menstrual Flow - Prolonged exposures to vibration could theoretically result in menstrual flow alterations. To date, female subjects at this facility have not reported any unusual problems, nor, to our knowledge, are any reported in the literature.

    e.   Breast Support - The forces experienced during vibration exposure under this protocol are relatively low. For your comfort, it is advised that appropriate breast support be used, similar to what might be used during exercise. Females with breast implants will not be allowed to participate in this study.

5. **Benefits:** You are not expected to benefit directly from participation in this research study, but a benefit to the Air Force is expected through an increased understanding of the effects of operational vibration on visual performance in these military flight environments.

6. **Compensation:** If you are active duty military you will receive your normal active duty pay.

7. **Alternatives:** Your alternative is to choose not to participate in this study. Refusal to participate will involve no penalty or loss of benefits to which you are otherwise entitled. You may discontinue participation at any time without penalty or loss of benefits to which you are otherwise entitled. Notify one of the investigators of this study to discontinue.

8. **Entitlements and confidentiality:**

    a.   Records of your participation in this study may only be disclosed according to federal law, including the Federal Privacy Act, 5 U.S.C. 552a, and its implementing regulations. Complete confidentiality cannot be promised, in particular for military personnel, whose health or fitness for duty information may be required to be reported to appropriate medical or command authorities. If such information is to be reported, you will be informed of what is being reported and the reason for the report.

    b.   You understand your entitlements to medical and dental care and/or compensation in the event of injury are governed by federal laws and regulations, and that if you desire further information you may contact the base legal office, 88 ABW/JA, DSN 787-6142. You may contact the medical consultant/ research monitor (Capt. Dawn Russell).

    c.   If an unanticipated event (medical misadventure) occurs during your participation in this study, you will be informed. If you are not competent at the time to understand the nature of the event, such information will be brought to the attention of your next of kin.

An Investigation and Analysis of the Vestibulo-ocular Reflex (VOR) in a Vibration Environment
FWR20130014H, Version 3.00
AFRL IRB Approval Valid from 15 Oct 2013 to 14 Oct 2014

Next of kin or emergency contact information:

Name_____   Phone#_____

d. The decision to participate in this research is completely voluntary on your part. No one may coerce or intimidate you into participating in this program. You are participating because you want to. Dr. Michael Miller, or an associate, has adequately answered any and all questions you have about this study, your participation, and the procedures involved. Dr. Michael Miller can be reached at (937) 255-3636 x4651. Dr. Michael Miller or an associate will be available to answer any questions concerning procedures throughout this study. If significant new findings develop during the course of this research, which may relate to your decision to continue participation, you will be informed. Refusal to participate will involve no penalty or loss of benefits to which you are otherwise entitled. You may discontinue participation at any time without penalty or loss of benefits to which you are otherwise entitled. Notify one of the investigators of this study to discontinue. The investigator or research monitor of this study may terminate your participation in this study if she or he feels this to be in your best interest. If you have any questions or concerns about your participation in this study or your rights as a research subject, please contact Col William Butler at DSN 986 – 5436 or william.butler2@wpafb.af.mil.

e. Your participation in this study may be photographed, filmed or audio/videotaped. These recordings will be archived on CD or similar media, as well as on a limited-access government computer. **The the data acquired during your participation will be maintained indefinitely and may become permanent records in the Air Force Collaborative Biomechanics Data Network.** Storage of the data is required to permit further data analysis and to provide a database against which future algorithms for overcoming image blur on helmet-mounted displays can be assessed. You consent to the use of these media and data for training, data collection, publication, and presentation purposes. Any release of records of your participation in this study may only be disclosed according to federal law, including the Federal Privacy Act, 55 U.S.C. 552a, and its implementing regulations. This means personal information will not be released to unauthorized source without your permission. These recordings and data may be used for presentation or publication, with your signed permission. They will be stored in a locked cabinet in a room that is locked when not occupied. Only the investigators of this study will have access to these media.

f. YOU FULLY UNDERSTAND THAT YOU ARE MAKING A DECISION WHETHER OR NOT TO PARTICIPATE. YOUR SIGNATURE INDICATES THAT YOU HAVE DECIDED TO PARTICIPATE HAVING READ THE INFORMATION PROVIDED ABOVE.

Volunteer Signature_____**Date**_____

An Investigation and Analysis of the Vestibulo-ocular Reflex (VOR) in a Vibration Environment
FWR20130014H, Version 3.00
AFRL IRB Approval Valid from 15 Oct 2013 to 14 Oct 2014

Volunteer Name (printed)_____

Advising Investigator Signature _____**Date**_____

Investigator Name (printed)_____

Witness Signature_____**Date**_____

Witness Name (printed)_____

We may wish to present some of the video/audio recordings from this study at scientific conventions or use photographs in journal publications. If you consent to the use of your image for publication or presentation in a scientific or academic setting, please sign below.

Volunteer Signature_____**Date**_____

Privacy Act Statement

<u>Authority</u>: We are requesting disclosure of personal information, to include your Social Security Number. Researchers are authorized to collect personal information (including social security numbers) on research subjects under The Privacy Act-5 USC 552a, 10 USC 55, 10 USC 8013, 32 CFR Part 219, 45 CFR Part 46, and EO 9397, November 1943 (SSN).
<u>Purpose</u>: It is possible that latent risks or injuries inherent in this experiment will not be discovered until some time in the future. The purpose of collecting this information is to aid researchers in locating you at a future date if further disclosures are appropriate.
<u>Routine Uses</u>: Information (including name and SSN) may be furnished to Federal, State and local agencies for any uses published by the Air Force in the Federal Register, 52 FR 16431, to include, furtherance of the research involved with this study and to provide medical care.
<u>Disclosure</u>: Disclosure of the requested information is voluntary. No adverse action whatsoever will be taken against you, and no privilege will be denied you based on the fact you do not disclose this information. However, your participation in this study may be impacted by a refusal to provide this information.
_____
ICD Distribution: Original filed with protocol records by PI; copy 1, subjec

An Investigation and Analysis of the Vestibulo-ocular Reflex (VOR) in a Vibration Environment
FWR20130014H, Version 3.00
AFRL IRB Approval Valid from 15 Oct 2013 to 14 Oct 2014

Appendix C: Testing Checklist

An Investigation and Analysis of the Vestibulo-ocular Reflex (VOR) in a Vibration Environment
Protocol FWR20130014H, Version 3.00

Subject #: _____

Session #: _____

Date: _____

Acceleration Level -   0.1 g   0.2 g

Video ON

Run	Freq (Hz)	Disp	Task	EOG Data	Accel Data	Video Data
1	0		Calib			
2	0		A			
3	2		A			
4	4		A			
5	6		A			
6	8		A			
7	10		A			
TWO MINUTE BREAK						
8	0		B			
9	2		B			
10	4		B			
11	6		B			
12	8		B			
13	10		B			
TWO MINUTE BREAK						
14	0		C			
15	2		C			
16	4		C			
17	6		C			
18	8		C			
19	10		C			

Video OFF

Appendix D: Post-Session Questionnaire

Post-Test Questionnaire

Subject ID:

Session #:

HEALTH

Are you experiencing any vestibular issues (dizziness, disorientation, nausea)?   YES / NO

Are you currently experiencing:

… pain in the musculoskeletal system especially in the back or neck?	YES / NO
… Numbness/Tingling/Weakness in Extremities?	YES / NO
… Headaches?	YES / NO
… Shooting Pain into Arms/Hands/Legs/Feet?	YES / NO

If any are YES, please report immediately to an investigator and/or medical monitor

TEST

How well did the helmet fit?   LOOSE / FINE / TIGHT

Next to each of the three tasks RATE from 1-10 (1 being easy, 10 being very difficult), the difficulty of the task.

Task A (Single Target Fixation):   _____

Task B (Smooth Target Tracking): _____

Task C (Jumping Target):   _____

Also, in the space provided, please write your overall impression of <u>each of</u> the tasks (what made it difficult, easy, boring, etc).

Task A (Single Target Fixation):

Task B (Smooth Target Tracking):

Task C (Jumping Target):

List in order from GREATEST to LEAST, the amount of vibration you experienced in each of these body parts? (Head, Upper Back/Chest, Lower Back/Abdomen, Buttocks, Upper Leg, Lower Legs/Feet)

_____

_____

_____   Amount of
                                  Vibration
_____   Experienced

_____

_____

Finally, please provide your overall thoughts/impressions of this experimental session. Please be as detailed as possible.

Appendix E: PSD Analysis

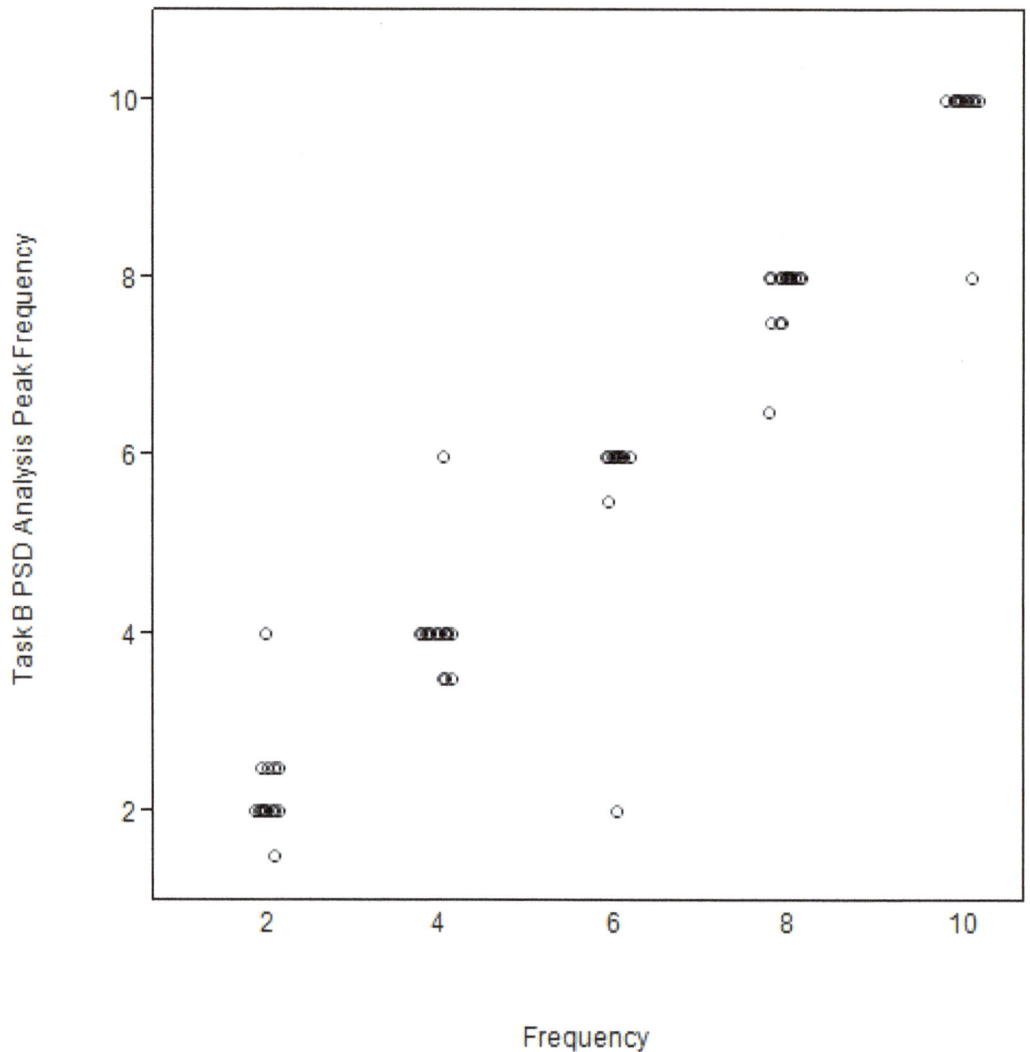

Figure E.1. Peak frequency from PSD analysis with corresponding input frequency.

Appendix F: Video Offset MATLAB Code

```
clear all
clc

% Video Analysis Tool
% It may help to use Ctrl+Enter on each segment

%% Import Video Object

    % give the video filename, and location(if not in MATLAB directory)
    file='L:\Research\HMD\Uribe_Thesis\Data\Raw↙
Data\EyeRecordings_wTimeSheets\Subject2\Subject2_Test1_Part1.mov';
    vidObj = VideoReader(file);  % Video Object
    vidHeight = vidObj.Height;  % Vert pixels
    vidWidth = vidObj.Width;  % Horizontal pixels
    nFrames = vidObj.NumberOfFrames; % Number of frames in video
    framerate = vidObj.Framerate;  % Framerate

%Recording Start    12:50:32    13:06:02
%
%Task A      12:51:02    12:54:32

%0 Hz   12:51:22    12:51:37
%2 Hz   12:51:57    12:52:12
%4 Hz   12:52:32    12:52:47
%6 Hz   12:53:07    12:53:22
%8 Hz   12:53:42    12:53:57
%10 Hz  12:54:17    12:54:32

    elap=etime( datevec('12:51:22',13),datevec('12:50:32',13));
    i0start= elap*framerate;
    elap=etime( datevec('12:51:37',13),datevec('12:50:32',13));
    i0end= elap*framerate;
    i0=[i0start i0end];

    elap=etime( datevec('12:51:57',13),datevec('12:50:32',13));
    i2start= elap*framerate;
    elap=etime( datevec('12:52:12',13),datevec('12:50:32',13));
    i2end= elap*framerate;
    i2=[i2start i2end];

    elap=etime( datevec('12:52:32',13),datevec('12:50:32',13));
    i4start= elap*framerate;
    elap=etime( datevec('12:52:47',13),datevec('12:50:32',13));
    i4end= elap*framerate;
    i4=[i4start i4end];

     elap=etime( datevec('12:53:07',13),datevec('12:50:32',13));
     i6start= elap*framerate;
    elap=etime( datevec('12:53:22',13),datevec('12:50:32',13));
    i6end= elap*framerate;
    i6=[i6start i6end];
```

```
    elap=etime( datevec('12:53:42',13),datevec('12:50:32',13));
    i8start= elap*framerate;
    elap=etime( datevec('12:53:57',13),datevec('12:50:32',13));
    i8end= elap*framerate;
    i8=[i8start i8end];

    elap=etime( datevec('12:54:17',13),datevec('12:50:32',13));
    i10start= elap*framerate;
    elap=etime( datevec('12:54:32',13),datevec('12:50:32',13));
    i10end= elap*framerate;
    i10=[i10start i10end];

    beg=i0start;
    begend=i0end;
    End=i10end;
    %End=i6end;

    % Preallocate variables
    mov(1:(End-beg)) = struct('cdata', zeros(vidHeight, vidWidth, 3,↙
'uint8'),'colormap', []);  % Preallocate movie structure (grayscale)
    mov(1).cdata = read(vidObj, beg);
    Standard = double(rgb2gray(mov(1).cdata));
    Comp=Standard(50:149,70:200);

    x=mymatchVertical(Standard,Comp,50:149,70:200);

%%  %%Correlation

for i=beg:End
    mov(i-(beg-1)).cdata = read(vidObj, i); % Read one frame at a time.
    movGray = double(rgb2gray(mov(i-(beg-1)).cdata));  % Grayscale image
    movGray=movGray(50:149,70:200);
    xpeak(i-(beg-1))=mymatchVertical(Standard,movGray,50:149,70:200);
end

movement=xpeak-x;

figure
subplot(6,1,1)
plot(movement(1:i0(2)-i0(1)))
title('0 Hz');
subplot(6,1,2)
plot(movement(i2(1)-i0(1)-100:i2(2)-i0(1)))
title('2 Hz');
subplot(6,1,3)
plot(movement(i4(1)-i0(1)-100:i4(2)-i0(1)))
title('4 Hz');
subplot(6,1,4)
plot(movement(i6(1)-i0(1)-100:i6(2)-i0(1)))
title('6 Hz');
```

```
subplot(6,1,5)
plot(movement(i8(1)-i0(1)-100:i8(2)-i0(1)))
title('8 Hz');
subplot(6,1,6)
plot(movement(i10(1)-i0(1)-100:i10(2)-i0(1)))
title('10 Hz');
```

```
function [ whichshift,location] = mymatchVertical(g, g1, Dx, Dy)
%UNTITILED Compares cropped image to new image 10 pixels up and down
%          in order to determine movement in new image
%   g is the full image
%   g1 smaller calibration image
%   Dx,Dy dimensions and location of the smaller image
location=zeros(21,1);

    for i=-10:10
        location(11+i)=(sum(sum(g1.*g(Dx+i,Dy))))/sqrt(sum(dot(g(Dx+i,Dy),g(Dx+i,Dy)))↙
*sum(dot(g1,g1)));
    end

    [x,whichshift]=max(location);
    whichshift=whichshift-11;

end
```

Appendix G: Fixation Task EOG Data Processing MATLAB Code

```matlab
%Obtain RMS and PSD values from EOG signal for fixation task
clear
close all

Sub=input('Subject number: ','s');
Sess=input('Session number: ','s');
fname=['L:\Research\HMD\Tung\Data Collection\Subject ' Sub '\Session ' Sess '\S' Sub↙
'_Session' Sess '_Calib.xls'];
data =xlsread(fname);

window = 500;
mask = ones(1, window)/window;

mov_avg= conv(data(2000:4000,1), mask, 'same');
Max=prctile(mov_avg,90);
mov_avg= conv(data(4000:6000,1), mask, 'same');
Min=prctile(mov_avg,10);
Dif=(Max-Min)/2;

load('L:\Research\HMD\MATLAB files\HighPass.mat')
b=unique(grpdelay(Hhp));

for k=2:6
    Freq=num2str((k-1)*2);
    fname=['L:\Research\HMD\Tung\Data Collection\Subject ' Sub '\Session ' Sess '\Task↙
A\S' Sub '_Session' Sess '_' Freq 'A.xls'];
    data1=xlsread(fname);
    figure
    plot(data1(:,1))
    name=['Subject ' Sub ' - Session ' Sess ' - ' Freq ' Hz - Task A'];
    title(name)

    data_vert=data1(:,1);

    Num_Blinks=input('How many blinks? ');

    if Num_Blinks>0
        for j=1:Num_Blinks
            [max_cc, locmax]= max(data_vert);

            if locmax < length(data_vert)-350
                data_vert=[data_vert(1:locmax-350);data_vert(locmax+350:end)];
            else
                data_vert=data_vert(1:locmax-350);
            end
        end
    end

    Vert=filter(Hhp,data_vert);
    Vert=Vert(b:end);
```

```matlab
%    Beg=input('Starting point: ');
%    End=input('Ending point: ');
%    data_vert=data1(Beg:End,1);

%    Mov_Avg = conv(data_vert(:,1), mask, 'same');
%    Mov_Avg_Vert = data_vert(:,1)- Mov_Avg;
    Vert=Vert/Dif;

    f1 = 1000;          %Sampling rate-Hz
    dCutoff = 150;      %Frequency cutoff-Hz
    f2 = f1*2;

    iNumPoints = ceil(dCutoff / 0.5) + 1;
    fr = (0:iNumPoints)' * 0.5;
    my_rms(1:iNumPoints+1,1) = fr;

    figure
    plot(Vert)
    title(name)

    %Start=input('Starting point (seconds): ');

    ps = pwelch(Vert, f2, f1, f2, f1);
    my_rms(1:iNumPoints,2) = sqrt(ps(1:iNumPoints) * f1 / f2);

    figure;
    plot(my_rms(1:iNumPoints+1,1),my_rms(1:iNumPoints+1,2))
    xlim([0 12]); ylim([0,.1]);
    xlabel('Frequency [Hz]'); ylabel('RMS');
    title('RMS vs Frequency for Vertical EOG');

    %Find and ouput RMS information for sample
    [x_lim,y_lim]=find(my_rms(:,1)==12);
    [value] = find(my_rms(:,1) == (k-1)*2);
    [v,i]= max(my_rms(4:x_lim,2));
    RMS_Freq(k,1) = my_rms(value(1),2);
    Max_RMS(k,1) = my_rms(i+3,2);
    Freq_max(k,1)= my_rms(i+3,1);

end
```

Appendix H: Tracking Task EOG Data Processing MATLAB Code

```matlab
close all

Sub=input('Subject number: ','s');
Sess=input('Session number: ','s');
fname=['L:\Research\HMD\Tung\Data Collection\Subject ' Sub '\Session ' Sess '\S' Sub↙
'_Session' Sess '_Calib.xls'];
data =xlsread(fname);

window = 500;
mask = ones(1, window)/window;

mov_avg= conv(data(2000:4000,1), mask, 'same');
Max=prctile(mov_avg,90);
mov_avg= conv(data(4000:6000,1), mask, 'same');
Min=prctile(mov_avg,10);
Dif=(Max-Min)/2;


load('L:\Research\HMD\MATLAB files\HighPass.mat')
b=unique(grpdelay(Hhp));

for k=2:6
    Freq=num2str((k-1)*2);
    fname=['L:\Research\HMD\Tung\Data Collection\Subject ' Sub '\Session ' Sess '\Task↙
B\S' Sub '_Session' Sess '_' Freq 'B.xls'];
    data=xlsread(fname);
    figure
    plot(data(:,1))
    name=['Subject ' Sub ' - Session ' Sess ' - ' Freq ' Hz - Task B'];
    title(name)
    hold

    Min=min(data(:,1));
    Max=max(data(:,1));

    line([2000 2000],[Min Max])
    line([3250 3250],[Min Max])
    line([4500 4500],[Min Max])
    line([7000 7000],[Min Max])
    line([9500 9500],[Min Max])
    line([12000 12000],[Min Max])
    line([13250 13250],[Min Max])
    line([14500 14500],[Min Max])

    data_vert=data(:,1);

    f1 = 1000;          %Sampling rate-Hz
    dCutoff = 150;      %Frequency cutoff-Hz
    f2 = f1*2;

    iNumPoints = ceil(dCutoff / 0.5) + 1;
```

```matlab
fr = (0:iNumPoints)' * 0.5;
my_rms(1:iNumPoints+1,1) = fr;

TopGood=input('Is top section good? (1=Yes, 0=No) ');
BotGood=input('Is bottom section good? (1=Yes, 0=No) ');

if TopGood==1 && BotGood==1
    Beg(1)=4500;
    End(1)=7000;
    Beg(2)=9500;
    End(2)=12000;

    for j=1:2
        Mov_Avg = conv(data(Beg(j):End(j),1), mask, 'same');
        Mov_Avg_Vert1 = data(Beg(j):End(j),1)- Mov_Avg;
        Mov_Avg_Vert=Mov_Avg_Vert1(250:(length(Mov_Avg_Vert1)-250));
        Vert=Mov_Avg_Vert/Dif;

        figure
        plot(Vert)
        name=['Subject ' Sub ' - Session ' Sess ' - ' Freq ' Hz - Task B ' j];
        title(name)
        xlim([0 2000]);

        ps = pwelch(Vert, f2, f1, f2, f1);
        my_rms(1:iNumPoints,2) = sqrt(ps(1:iNumPoints) * f1 / f2);

        figure;
        plot(my_rms(1:iNumPoints+1,1),my_rms(1:iNumPoints+1,2))
        xlim([0 12]); ylim([0,.1]);
        xlabel('Frequency [Hz]'); ylabel('RMS');
        title('RMS vs Frequency for Vertical EOG');

        %RMS2(j,i)=rms(Vert);

        [x_lim,y_lim]=find(my_rms(:,1)==12);
        [value] = find(my_rms(:,1) == (k-1)*2);
        [v,i]= max(my_rms(4:x_lim,2));
        RMS_Freq2(j,1) = my_rms(value(1),2);
        Max_RMS(j,1) = my_rms(i+3,2);
        Freq_max(j,1)= my_rms(i+3,1);

        %          figure
        %          pwelch(Vert, 500, [], length(Vert), 1000,'onesided');
        %          title(name)
        %          xlim([0 20]);
    end
    RMS(1,k)=mean(RMS2(:,k));
else
    if TopGood==1
        Beg1=4500;
```

```matlab
            End1=7000;
        else
            Beg1=9500;
            End1=12000;
        end
        Vert=filter(Hhp,data_vert);
        Vert=Vert(Beg1+b:End1+b);
        Vert=Vert/Dif;

        figure
        plot(Vert)
        name=['Subject ' Sub ' - Session ' Sess ' - ' Freq ' Hz - Task B ' j];
        title(name)
        xlim([0 2000]);

        ps = pwelch(Vert, f2, f1, f2, f1);
        my_rms(1:iNumPoints,2) = sqrt(ps(1:iNumPoints) * f1 / f2);

        figure;
        plot(my_rms(1:iNumPoints+1,1),my_rms(1:iNumPoints+1,2))
        xlim([0 12]); ylim([0,.1]);
        xlabel('Frequency [Hz]'); ylabel('RMS');
        title('RMS vs Frequency for Vertical EOG');

        %RMS(1,i)=rms(Vert);

        [x_lim,y_lim]=find(my_rms(:,1)==12);
        [value] = find(my_rms(:,1) == (k-1)*2);
        [v,i]= max(my_rms(4:x_lim,2));
        RMS_Freq(k,1) = my_rms(value(1),2);
        Max_RMS(k,1) = my_rms(i+3,2);
        Freq_max(k,1)= my_rms(i+3,1);
    end

end
```

Appendix I: High Pass Filter

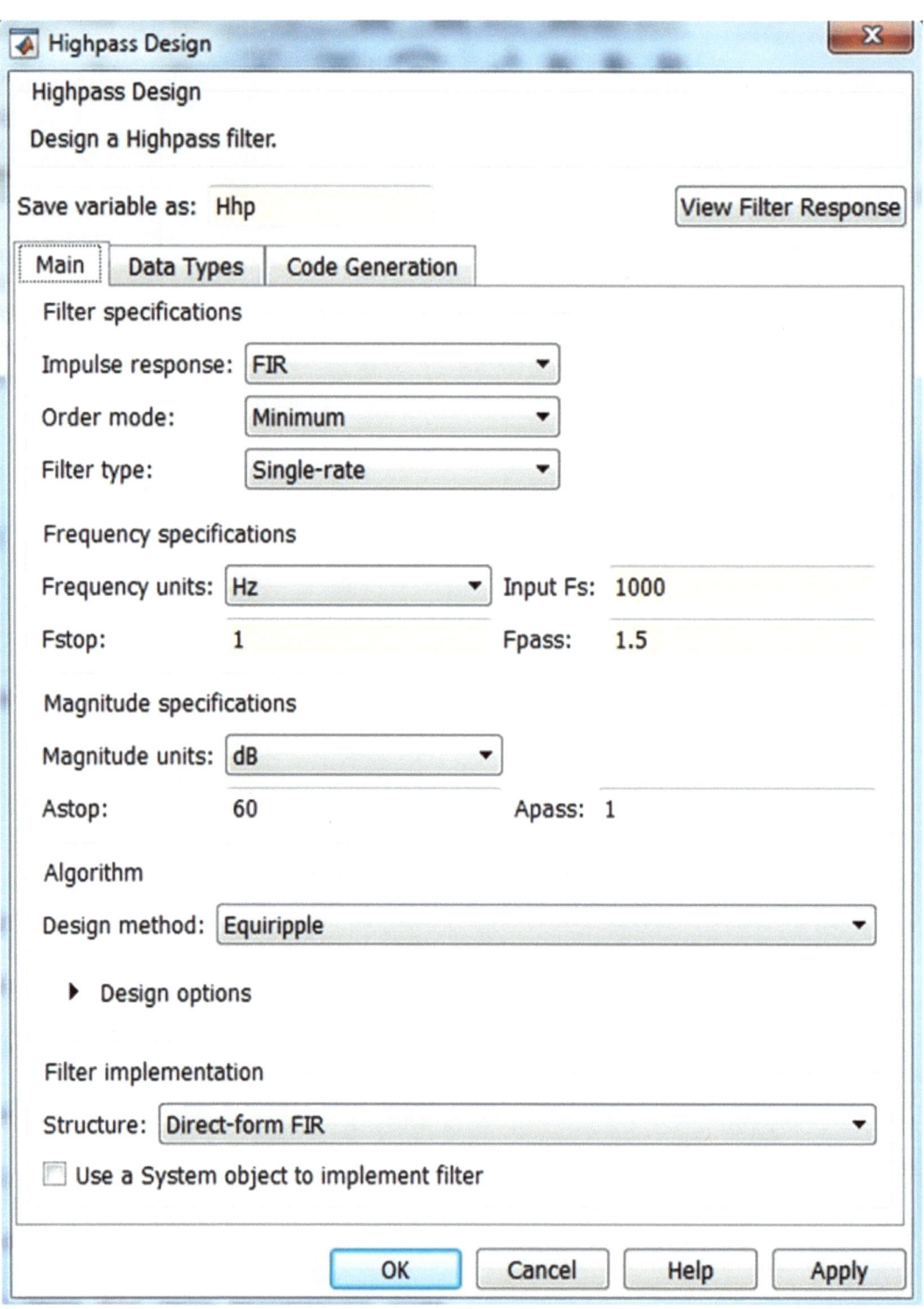

Figure I.1. Inputs into the filterbuilder function in MATLAB to create the high pass filter used in the EOG signal analysis.

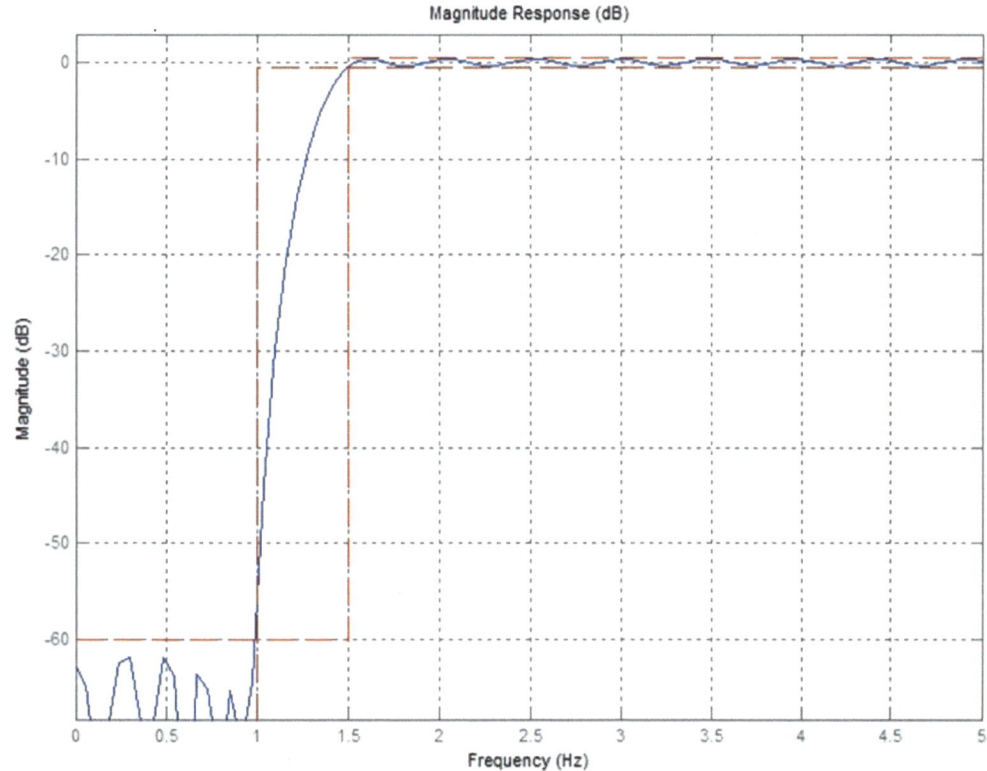

Figure I.2. Filter response in the frequency domain of high pass filter created in MATLAB.